I0797950

BISON
BOOKS

# DINOSAUR DREAMS

## A Father and Daughter in Search of America's Prehistoric Past

B. J. HOLLARS

University of Nebraska Press
*Lincoln*

Manufactured in the United States of America

The University of Nebraska Press is part of a land-grant institution with campuses and programs on the past, present, and future homelands of the Pawnee, Ponca, Otoe-Missouria, Omaha, Dakota, Lakota, Kaw, Cheyenne, and Arapaho Peoples, as well as those of the relocated Ho-Chunk, Sac and Fox, and Iowa Peoples.

Publication of this volume was assisted by the University of Wisconsin–Eau Claire's Office of Research and Sponsored Programs and Department of English.

For customers in the EU with safety/GPSR concerns, contact:
gpsr@mare-nostrum.co.uk
Mare Nostrum Group BV
Mauritskade 21D
1091 GC Amsterdam
The Netherlands

Library of Congress Control Number: 2024061767

Designed and set in Arno Pro by Lacey Losh.

To Ellie—strong as an *Ankylosaurus,* gentle as a *Maiasaura.*

“Paleontology is one of the few remaining sciences in which amateurs still make contributions. Equipped with a shovel, pick, and a little knowledge, almost anyone can mount her own search for fossils.”

—*Dinosaur Lives*, by John R. Horner and Edwin Dobb

# Contents

# Illustrations

# Acknowledgments

I owe a dinosaur-sized thank you to so many, beginning with the friends we met along the way: Tom Chase, Kerry Libby, Zach Perry, Tim Busby, Jacque Gregg, Tom Hebert, Sean Doyle, Susan Luinstra, Dave Trexler, Lila and Dan Redding, Lela Patera, Samantha French, Dixie Stordahl, Lori Taylor, Sue Dalbey, Dave Fuqua, Riley Bell, Sue Veroye, Tom Shoush, Robert Canen, Dr. Nathan Carroll, Pat Rouane, Maxx Martel, James Kelly, Brendan Heidner, and Dr. Jack Horner.

Thanks, too, to Victor Bjornberg, Clayton Phipps, Adrienne Mayor, Dr. Sabre Moore, and Kallie Moore, who provided additional context and stories.

Thank you to my inspiring students at the University of Wisconsin–Eau Claire—too many of you to name. To my colleagues and friends at the University of Wisconsin–Eau Claire: Chancellor Jim Schmidt, Interim Provost Mike Carney, Foundation Executive Director Kimera Way, Foundation President Curt Krizan, Foundation Vice President Julia Diggins, Dean Aleks Sternfeld-Dunn, Dean Carmen Manning, Jon Loomis, Allyson Loomis, Molly Patterson, Brett Beach, Dr. Dorothy Chan, Dr. Asha Sen, Dr. Sarita Mizin, Dr. Kaia Simon, Dr. Jonathan Rylander, Dr. David Jones, Dr. David Shih, Dr. José Alvergue, Dr. Stephanie Farrar, Dr. Stephanie Turner, Dr. Joel Pace, Dr. Cathy Rex, Dr. Stacy Thompson, Dr. Blake Westerlund, Dr. Lynsey Wolter, Dr. Shanna Cameron, Dr. Aubrie Warner, Dr. Sean Weidman, Dr. Heather Fielding, Dr. Laura Jok, Dr. Jennifer Ervin, Amy Fleury, Shelley Donnelly, Dr. Sara Monahan, Jacob Boyd, Joseph Serio, Isabella Gross, Christy Potvin, Kate Worzala, Greg Kocken, Dr. Justin Patchin, Dr. Jason Spraitz, Dr. Paul Thomas, Dr. Jeff DeGrave, Professor Arthur Grothe, Dr. Chiayu Hsu, and especially Candis Sessions.

Thanks to Dr. Erica Benson and the Office of Research and Sponsored Programs at the University of Wisconsin–Eau Claire, whose support proved vital to this project, and to the University of Wisconsin–Eau Claire Academic Affairs Professional Development Program, which also provided support.

To the around-town crew: the late Jim Alf, Nick Butler, John Hildebrand, Max Garland, Nick Meyer, Brian Macki, Tom Giffey, McKenna Scherer, Jim McDougall, Kyran Hamill, Bruce Taylor, Patti See, the late Brady Foust, Sarah Johnson, and the regional writing community. Thank you to Julian Emerson and Andy Patrie. And to Ken Szymanski and Eric Rasmussen, early and often readers of this book. To Dathan Boardman, thank you for sharing your art in support of my own. To Brendan Todt, for the correspondence and the kinship. To the Dayton family for kindness and good cheer. To Mom, Dad, Brian, and Janae. To Emily Taylor and the early morning crew—never suffer alone.

To all my friends at Bison Books and the University of Nebraska Press, especially Clark Whitehorn.

To Meredith, Henry, and Millie. And Ellie most of all.

Finally, and importantly, this work was conducted on the ancestral land of the Assiniboine, Blackfeet, Chippewa, Cree, Crow, Gros Ventre, Kootenai, Little Shell Chippewa, Northern Cheyenne, Pend d'Oreille, Chippewa Cree, Salish, and Sioux Indigenous nations. We are grateful for the opportunity to have learned and lived upon it for a time.

# Author's Note

Dinosaurs are complex creatures, and attempting to write a book about them has made for a complex task. However, that task was made simpler thanks to the many experts who supported us throughout our adventure. I am indebted to those experts, all named in the acknowledgments. I am equally indebted to the acquaintances we met along the way. These interactions were hardly planned but important to our understanding of the world and our place within it.

My primary sources came in many forms, though mainly interviews. All passages within quotation marks should be considered authentic dialogue, either recorded or remembered. For clarity and readability purposes, I have occasionally tweaked an interviewee's response while remaining mindful to preserve its meaning. It's worth noting that my number one interviewee wasn't an interviewee at all but my daughter and sidekick, nine-year-old Ellie. While I like to think our banter is always as delightful as depicted herein, a time or two, I claimed parental privilege and enhanced our delightfulness just a smidgen.

The bibliography at the end of this book contains a complete list of sources. For now, I'll thank the authors I relied on most frequently: Stephen L. Brusatte, Jack Horner, Adrienne Mayor, Charles Sternberg, and David Rains Wallace.

When working with *The Journals of Lewis and Clark*, I occasionally updated the spelling of key words for readability purposes. These changes are bracketed. The original entries (including the original spellings) can be viewed at the journal's online repository at https://lewisandclarkjournals.unl.edu/.

When writing about America's first inhabitants, I made every effort to reference specific tribal nations. Additionally, upon initially referencing a specific battle from the American Indian Wars, I offer the Americanized and Indigenous battle names.

Finally, to dinosaur lovers everywhere, I beg your pardon in advance. Given the breadth and depth of timelines and geographies, this book falls far short of a comprehensive history of America's prehistoric past. Rather, it's an entry point into the larger story—a foray into Montana's Mesozoic era and back again, with a father and daughter as guides.

# Cast of Characters

**Richard Baker** is a musician with the Rivertown Rounders in Great Falls, Montana.

**Robert Bakker** is an American paleontologist who contributed to the Dinosaur Renaissance by supporting the idea that some dinosaurs were warm-blooded. He was a protégé of American paleontologist John Ostrom.

**Victor Bjornberg** served as the Montana Office of Tourism's tourism development and education coordinator during the creation of the Montana Dinosaur Trail. Today, he serves as the trail's volunteer coordinator.

**Marion Brandvold** (1912–2014) of Choteau, Montana, was a lifelong amateur paleontologist credited with finding the first baby dinosaur bones. Her discovery made possible Jack Horner and Bob Makela's identification of *Maiasaura* and their theory that some dinosaur species cared for their young. She is the mother of paleontologist Dave Trexler.

**Barnum Brown** (1873–1963) was an American paleontologist who famously unearthed the first documented remains of a *Tyrannosaurus rex* in 1902 near modern-day Jordan, Montana. Named after P. T. Barnum (and a showman himself), Brown also served as a paleontological promoter to the American public throughout his life.

**Tim Busby** is the director of the Upper Musselshell Museum in Harlowton, Montana.

**Robert Canen** is the director of the Glendive Dinosaur and Fossil Museum, which offers a creationist perspective on the planet's age and formation.

**William Clark** (1770–1838) was an American explorer and territorial governor best remembered as one-half of the leadership team of the Lewis and Clark Expedition.

**Edward Drinker Cope** (1840–97) was an American paleontologist best remembered for participating in the Bone Wars against rival Othniel Charles Marsh. At nineteen, Cope published his first scientific paper, fueling a lifetime of paleontological discovery. He is often credited with naming at least fifty-six dinosaur species.

**Sue Dalbey** is a natural resource specialist at the Fort Peck Interpretive Center in Fort Peck, Montana.

**Sean Doyle** is the director of the Old Trail Museum in Choteau, Montana.

**Samantha French** is the director of the Blaine County Museum in Chinook, Montana.

**Dave Fuqua** is the YouTube star of Dinodave Paleo Adventures and a native of Glendive, Montana.

**Diane Gabriel** (1946–94) was a paleontologist from Museum of the Rockies who conducted several digs in Makoshika State Park. Today, the park has a trail named in her honor.

**Jacque Gregg** is a volunteer at the Garfield County Museum in Jordan, Montana.

**Tom Hebert** is the director of Earth Sciences Foundation Inc.

**Brendan Heidner** is a reporter for the Glendive newspaper.

**Jack Horner** is a renowned American paleontologist best known for describing the *Maiasaura* and furthering the case that some dinosaurs cared for their young. Horner served as a technical director for several *Jurassic Park* films and is often credited as the inspiration for Dr. Alan Grant. Horner previously served as the curator of paleontology at the Museum of the Rockies and teaches at Chapman University.

**Thomas Jefferson** (1743–1826) served as the third president of the United States and, in that capacity, dispatched Captain Meriwether Lewis and Lieutenant William Clark to explore the newly purchased Louisiana Purchase. Given his deep love of paleontology, he encouraged Lewis and Clark and their Corps of Discovery to look for ancient fauna, most notably mastodons (which he often called mammoths), which he believed might still have been alive in the early nineteenth century.

**James Kelly** is a motorcyclist from eastern Washington.

**Marshall Lambert** (1914–2005) was an amateur paleontologist and teacher who served as the director of the Carter County Museum in Ekalaka, Montana, for fifty years.

**Meriwether Lewis** (1774–1809) was an American explorer and politician best remembered as one-half of the leadership team of the Lewis and Clark Expedition.

**Kerry Libby** is former external affairs director and current executive director for the Standing Rock Sioux Tribe.

**Susan Luinstra** taught at Bynum School in Bynum, Montana, for thirty-eight years. In 2007 she was named the National Rural Education Association's Teacher of the Year. In 2017 she was featured on NBC's *TODAY* for providing students with morning dance lessons.

**Othniel Charles Marsh** (1831–99) was an American paleontologist most remembered for participating in the Bone Wars against rival Edward Drinker Cope. Marsh's work helped lay the groundwork for the dinosaur collections in the Peabody Museum of Natural History (named for Marsh's uncle, George Peabody) and the Smithsonian Institution. He is often credited with naming eighty or more dinosaur species.

**Maxx Martel** is a private collector of fossils and historical artifacts in Glendive, Montana.

**Josey Montana** is a server (and ghost story aficionado) at the Grand Hotel and Restaurant in Big Timber, Montana.

**John Ostrom** (1928–2005) was an American paleontologist who theorized that birds were the closest living relatives of dinosaurs. His work changed scientists' perception of dinosaurs from being cold-blooded reptiles to—in some species—more akin to warm-blooded birds.

**Sir Richard Owen** (1804–92) was an English biologist who, in 1842, coined the phrase *Dinosauria*.

**Lela Patera** is a board member and volunteer for the H. Earl Clack Museum in Havre, Montana.

**Zach Perry** is the assistant store manager at Museum of the Rockies and a paleontology graduate student at Montana State University.

**Clayton Phipps**, also known as "The Dinosaur Cowboy," is a commercial fossil hunter most famous for his 2006 discovery of the Dueling Dinosaurs. He is also the star of Discovery Channel's reality show *Dino Hunters*.

**Lila Redding** is a longtime volunteer and founding member of the Depot Museum in Rudyard, Montana.

**Pat Rouane** is the gift shop manager at the Carter County Museum in Ekalaka, Montana, and a graduate of Montana State University's Department of Earth Sciences program.

**Charles Sternberg** (1850–1943) was an American paleontologist most famous for discovering the "*Trachodon* mummy" (an *Edmontosaurus* specimen) and serving as an assistant to Edward Drinker Cope. Sternberg's book *The Life of a Fossil Hunter* provides a unique firsthand account of late nineteenth- and early twentieth-century paleontology in America.

**Diane Stinger** is a musician with the Rivertown Rounders in Great Falls, Montana.

**Adrienne Swallow** (1958–2015) founded the Standing Rock Sioux Tribe's Paleontology Department and was instrumental in forming the Standing Rock Institute of Natural History.

**Dave Trexler** is an American paleontologist and founder of the Two Medicine Dinosaur Center (also known as the Montana Dinosaur Center). In 1978 he and his mother, Marion, discovered juvenile *Hadrosaur* fossils. More recently, Trexler also published the first paper on and later helped rejacket Leonardo, the dinosaur mummy, for transport to NASA's Johnson Space Center in Houston. Additionally, he is the author of *Becoming Dinosaurs: A Prehistoric Perspective on Climate Change Today*.

# DINOSAUR DREAMS

FIG. 1. Montana Dinosaur Trail Map. Courtesy The Montana Dinosaur Trail.

FIG. 2. A father and daughter pose with a *Triceratops* model in Rapid City, South Dakota's Dinosaur Park. Sternberg, George Fryer 1883–1969, "064_05: Dinosaur Model Pose Wooster and Daughter" (2021). *George Sternberg Album #4*. 297, Forsyth Library Special Collections, Fort Hays State University.

# Prologue

The bad news is that we're doomed.

The good news is that we've always been doomed.

And the best news of all is that we're not doomed yet.

The same was not true for the dinosaurs sixty-six million years ago, when an asteroid nearly seven miles wide crashed into what would become the Yucatán Peninsula. One moment, the dinosaurs were minding their own business—roaming the forests, swamps, and deserts—and the next, they peered up to see the beginning of their end hurtling toward them at forty-five thousand miles per hour. Upon the asteroid's impact, the world caught fire. Molten material set the forests aflame. Tsunamis swept the shorelines, rocks rose and fell, and plumes of debris scattered across the atmosphere. The explosive energy released has been compared to "4.5 billion times the explosive power of the Hiroshima atomic bomb"—a measure that is all but meaningless given its enormity.

But according to research from 2013, it wasn't the asteroid that did the dinosaurs in—it was the long-term impact of the blast. Vaporized material and soot combined to form a cloud, coating the earth, blotting out the sun, and halting nearly all photosynthetic activity. Phytoplankton in the oceans died, prompting a decline in oxygen in the atmosphere. Plants withered; plant-eaters perished. The population of carnivores soon collapsed. A global winter frosted the land, and over the next thirty-three thousand years—amid this swiftly changing world—75 percent of all species discovered that the only home they'd ever known had suddenly become uninhabitable.

What does this have to do with humans? For starters, we wouldn't be here if the Cretaceous-Paleogene extinction event had not occurred. When the all-consuming dinosaurs died, more diminutive species suddenly found themselves in a position to flourish—everything from small mammals, lizards, turtles, and salamanders to crocodiles, alligators, and birds. For a geological moment, the meek really had inherited the earth. And many of them were destined to make good on their inheritance.

Fast-forward to seven million years ago to witness the emergence of our earliest human ancestors, an extinct genus of hominid called *Sahelanthropus*. Fast-forward to three hundred thousand years ago, and *Homo sapiens* entered the scene. As a result of all that had come before (the asteroid, the fires, the soot), we humans were finally given our shot. A shot we might've missed altogether had the dinosaurs stuck around. Food chains had to change, as did the competition for resources. Imagine the seemingly endless biological, chemical, and atmospheric factors that coalesced to allow us to be here at all.

One might think we'd be humbled by the notion that we owe our existence to an asteroid that wiped the world clean. Or we might acknowledge that past events have always led to future events. Or that in recent history, human behavior has played an outsized role in shaping some rather foreboding future events.

Anthropologist Loren Eiseley popularized a line from poet Francis Thompson that speaks to humankind's deep connection to our world: "One could not pluck a flower without troubling a star."

Humans have plucked many flowers. We've troubled many stars.

What remains to be seen is if it's possible to return those flowers to their roots, to set those stars back on their courses.

*

In a time before phrases like "heat dome" became part of the daily parlance, my daughter Ellie fell in love with dinosaurs. Her first encounter came through a song: Laurie Berkner's "We Are the Dinosaurs," a catchy tune we blasted in the van throughout Ellie's early years. Its lyrics encouraged listeners to behave as dinosaurs might—marching, eating, resting, and roaring; all fine ways for a child to pass the time while buckled into the back of a van.

Over the years, we played that song at least a thousand times. Her older brother Henry tired of it; her younger sister Millie did too. But never Ellie, for whom it became an anthem. Long after the song ended, long after the unbuckling, she could often be found roaming around the house, marching, eating, resting, and roaring. Teeth bared, claws gleaming, sometimes she was a stalking *Velociraptor*; other times, the armored *Ankylosaurus*. On a good day, the dog escaped unscathed. On a bad day, none of us did.

By age nine, many of Ellie's more dinosaurian tendencies had faded, replaced by the usual diversions—friends, sports, and screens. However, sometimes, on long drives, when feeling nostalgic, I'll cue up the tune and glance in the rearview to spot Ellie still wild-eyed and roaring.

"What do you love most about dinosaurs?" I once asked toddler Ellie.

"Umm . . ." she pondered, tapping her chin, "probably that they're dinosaurs!"

Ask a paleontologist that question, and you might receive some variation of the same answer. Regardless of age or expertise, we love dinosaurs for their very being. We love them because they once were where we stand today, thunderously roving amid a very different landscape. We love them for their mysteries, their marvels, their impossible size. We love them because, though it may appear as if we've lapped them on the evolutionary scale, they were some of the original evolutionary explorers—shedding their skin, again and again, to adapt to their changing world.

But there are also more nuanced reasons why we love dinosaurs. My love for the "terrible lizards" is rooted in my belief that they might have something to teach us. After all, no species has felt the full impact of a changed climate quite like dinosaurs, who endured vastly altered climactic conditions before they died. According to a 2019 *New Yorker* article, a trillion tons of carbon dioxide and ten billion tons of methane were introduced into the atmosphere following the asteroid strike. Today, we call those greenhouse gases. And we know, too, that their presence can warm the planet and imperil our ability to survive. What might dinosaurs teach us about our warming planet? What cautionary tales might be gleaned from their fossil record?

"The best way to look ahead is to look behind, at those organisms, including dinosaurs, that survived extended climate change," writes Mary H. Schweitzer, research curator of vertebrate paleontology at the North

Carolina Museum of Natural Sciences. "Only by understanding the geologic record of diversity, adaptation and climate variability," she argues, "can we hope to face the challenges ahead."

Challenges include rising temperatures at a rate not observed in ten thousand years, declining biodiversity that threatens one million species with extinction, and a human population boom that has quadrupled over the past century. Worst of all, we can't isolate our problems when all our problems are inextricably linked. The web of life is also the web of death. Pull a thread and the whole thing begins to unravel. Insert any metaphor you like—falling dominoes, stacking blocks, a house of cards—we are always at the center of the collapse.

As we enter the Anthropocene—the geological epoch whose primary characteristic is human activity's impact on the climate and environment—might it behoove us to take a page from the prehistoric playbook? To reacquaint ourselves with the literal end of an era from a distance of sixty-six million years?

Temporally, dinosaurs seem a long way off, but geographically, not so much.

There is no shortage of dinosaur fossils in the American West. At the right place, and with the right guide, you can throw a rock and hit a prehistoric fossil. (Not that I'd recommend it.) Instead, I'd encourage seeing those fossils firsthand, which Ellie and I did for two glorious weeks in the summer of 2023.

Specifically, we joined a nearly-century-long caravan of dinosaur tourists in America. While those who came before often settled for hit-or-miss dinosaur-themed roadside attractions, we embarked upon the Montana Dinosaur Trail, a clearly defined route of fourteen dino-themed stops in a state famous for its dinosaurs. As if the dinosaurs were not enough, MDT travelers are also equipped with a handy-dandy passport. If one completes the passport—earning a stamp at each of the fourteen museums, interpretive centers, and dig sites—the recipient is eligible to receive the greatest of honors: a free T-shirt.

There are few things I won't do for a free T-shirt. Dedicating two weeks to driving a 2,500-mile round trip between Wisconsin and Montana is just the latest in what I hope will be a lifetime of ill-advised, T-shirt-inspired adventures.

But our journey was about more than free apparel. It was about learning to live a full life in the shadow of a somewhat ecologically terrifying time.

Of course, the climate crisis is one of many scenarios that might lead us to our end. Nuclear annihilation is also on the table, as is the unbridled rise of artificial intelligence. Some days, it's hard to know which potential mass extinction event we should prioritize. Our whack-a-mole approach to engaging with these apocalyptic flavors of the week is unnerving, to say the least.

Yet all this doom and gloom is easier to bear alongside my daughter. This is why Ellie and I were near-perfect companions for the Montana Dinosaur Trail: I needed her fortitude, and she needed my driver's license.

But our partnership worked for another reason too.

Ellie is the Winnie the Pooh to my Eeyore, the SpongeBob to my Squidward, and the Ted Lasso to my Roy Kent. And she is hardly alone in her quixotic approach to the perils of the modern world. Like many nine-year-olds, she's kept up at night, sweating not the Climate Change Performance Index but whether she remembered to return her library book, how to spell "daffodil," and whether she'll get invited to Claire's birthday party.

Where I see fossils, she sees a *Brachiosaurus* chewing the juiciest leaves in the uppermost branches of the trees. Where I see an asteroid, she sees a *Maiasaura* bedding down for the night with its babies. Simply put, she offsets my doomsayer glasses with her rose-colored ones. When life gives her lemons, she takes a big, sour bite right out of the peel.

Though the previous pages might lead one to think otherwise, believe me when I tell you this is a hope-filled book. It's the story of a father, a daughter, and a full tank of gas. It's a story of seeking clarity and connection with the world and its people. And it's a story of how these two weeks changed us both, often unexpectedly.

Beyond all that, it's also the story of a series of small towns throughout Montana, each worthy of our time, attention, and tourism. And it's the story of the Montana Dinosaur Trail—now in its twentieth year—which links these primarily rural towns so folks like Ellie and me can receive a carefully curated, once-in-a-lifetime experience among dinosaurs.

Finally, it's a story of experts and amateurs: professional paleontologists (credentialed and degreed), amateur paleontologists (less credentialed and degreed), and commercial fossil hunters (who find fossils to make a sale). Collectively, theirs is a story of egos, ethics, paychecks, and privilege. I didn't intend to tackle this topic, though their differing perspectives on dinosaur fossils serve as a microcosm to the largest question of all:

Can we come together to extend our time on Earth despite humankind's vast differences?

One thing we know for sure is that humans are not easily humbled, though we ought to be. Whereas dinosaurs ruled the earth for nearly 165 million years, we *Homo sapiens* have so far barely managed three hundred thousand. Who could've dreamed we could cause such damage in such a short geologic timeframe? Or that humanity's most lasting "achievement" might be the speed with which we destroyed our planetary home?

One day, while doomscrolling on social media, I stumbled upon a photo of two park rangers posing alongside a digital display of the unofficial heat reading from July 16, 2023. The display read: "132° Fahrenheit, 55° Celsius." Despite the all-but-uninhabitable temperatures blazing across the display, the rangers smiled and playfully gesticulated toward the temperature readout. I might've kept scrolling had it not been for the accompanying tagline:

"This is like dinosaurs having their picture taken with an asteroid."

First, I felt sick. But then I felt inspired.

Maybe the song was right; maybe we are the dinosaurs.

But we're not the dinosaurs yet.

FIG. 3. B.J. and Ellie in the driveway of their Eau Claire, Wisconsin home moments before embarking on their trip to Montana. Author photo.

# 1
# A Bone to Pick

**JUNE 29**

*Eau Claire,* WI → *Bismarck,* ND

Because no asteroids are hurtling toward Earth, there's no need to rush our goodbye. Nine-year-old Ellie and I take our time, loitering alongside the minivan parked in our driveway on another warm morning in a series of warm mornings in the hottest June on record.

Eleven-year-old Henry and three-year-old Millie flank their mother Meredith, who is unsure how to proceed. Ellie and I are about to embark upon the adventure of a lifetime. Not another Thursday trip to the park, the library, or the grocery store, but Montana—the Treasure State—home to natural beauty, mineral resources, and, most of all, dinosaurs.

We're dedicating two weeks and 2,500 miles to complete the Montana Dinosaur Trail—a statewide trail of fourteen museums, interpretive centers, and dig sites—all dedicated to the mystery and majesty of the largest land animals ever to roam the earth. More than a few members of the Dinosaur Hall of Fame once called present-day Montana home, from the razor-toothed *Tyrannosaurus rex* to the duckbilled *Maiasaura*. Enter the *Triceratops*, the *Brachylophosaurus*, the *Pachycephalosaurus*, and more. Anyone who was anyone in the Late Cretaceous lived and died where we're now headed.

Meredith shoots me her bittersweet smile: there's no sense prolonging the inevitable.

"Okay then," I say, beginning the proceedings. "I guess it's now or never."

That isn't entirely true.

The dinosaurs aren't going anywhere. They're either safely displayed in museums or buried beneath the ground. But if we want to make Bismarck by nightfall, we need to hit the road. My anxiety about hitting that road is only offset by my anxiety about staying put, which would be much easier. But easy is not what inspires us. For the rest of our days, I want us to recall that time in our lives when we didn't do the easy thing. The time we threw caution to the wind and were rewarded (fingers crossed) with memories more numerous than stars.

When I broached the idea the previous fall, Ellie seemed thrilled. Dad and dinosaurs! What could be better than that? But as summer approached, she'd grown leery of the timeline and the company.

"Two weeks is a long time," she conceded.

"Not really," I replied, "not in a geological sense . . ."

"Well, it feels long," she added sheepishly, "especially when that whole time is spent with one person . . ."

"No offense taken," I said.

But I knew what she meant. Distance makes the heart grow fonder, though it can be hard to come by when cramped in a van.

Meredith makes the first move, embracing Ellie like a *Velociraptor* on a small mammal—but less murderous.

"Oh, I love you so much, my sweet girl," she says.

Ellie—who has never been separated from her mom for longer than a mini-week at camp—sags into her mother's open arms.

At that moment, I suspect Ellie might have preferred the park, library, or store over our two-week trip to Montana. What are dinosaurs, after all, but a bunch of old bones? Don't get me wrong, Ellie loves dinosaurs (she's a living, breathing child, after all). But she also loves her bunk bed, the backyard trampoline, and, often, her brother and sister.

But it's too late to turn back. Already, I'd spent the morning playing *Tetris* with our provisions, fitting coolers, sleeping bags, suitcases, and backpacks into their perfect places. So no, we are not turning back, and we are not opening the trunk. We're fifteen feet from our living room, but our journey has begun.

Besides, did dinosaurs throw in the towel when they missed their mommies? Did they pull their hybrid minivans back into their garages and return to the safety of their living rooms? They most certainly didn't. Come hell or high water (and they endured both), they persisted.

For 165 million years.

With a little luck and a lot of fast food, Ellie and I can survive two weeks together.

Taking his cue from his mom, Henry musters the most heartfelt goodbye an eleven-year-old can.

"Okay," he sighs, "see you later or . . . whatever."

"I love you, Henry," Ellie says, latching onto him.

Completing the sibling chain, Millie clings to Ellie's leg like a barnacle—an extraction that requires both parents to peel her fingers away.

"Bring her back safe," Meredith says, hugging me. "And find lots of dinosaurs!"

"Will do," I promise. "At least the first part."

The previous night, under the cover of darkness, the entire family had tried to slow time by chalk-drawing in the driveway. One after another, as the moon gleamed overhead, we lay flat on the blacktop so Millie could trace our outlines with thick nubs of chalk. No matter that our chalky shapes resembled a crime scene; for Millie, it was a way to keep us close.

Peering at our two-dimensional family in the morning light, I wonder how long our chalk versions will last. No doubt there's a downpour somewhere in our future, though for the moment, it's nothing but sunshine ahead. Millie—who seemed on the verge of collapse just moments ago—forgoes her freakout in favor of one last addition to her chalk drawing. Reaching for a piece of chalk, she draws a top hat along Ellie's chalked head.

"There," Millie says proudly, turning toward us. "Now, Ellie, you a gentleman!"

Laughing beats crying, so we do.

Ellie and I enter the van, buckle our seatbelts, and begin the long, slow reverse out of the driveway.

I catch Ellie's anxious eyes in the rearview.

"It'll be fine," I promise.

"Are you sure?" she asks.

"Sure, I'm sure," I say.

Which is probably what some daddy dinosaur said to his daughter as an asteroid lit up the sky.

*

Regarding things not being fine, I give you America in the 1860s. Amid the backdrop of the Civil War and the American Indian Wars, a third war was brewing: the Bone Wars. This late nineteenth-century battle royal was fought between Edward Drinker Cope and Othniel Charles Marsh, men whose passion for paleontology was dwarfed only by their vitriol for one another.

In one corner, from the great state of Pennsylvania, was Edward Drinker Cope—a high-performing scientific savant who never found a fossil he didn't want to name.

In the other, hailing from New York, was Othniel Charles Marsh—a brooding and calculating fellow from the hallowed halls of academia.

Had they found themselves in the ring together, Cope might've opted for the bob and weave, while Marsh would've preferred the hard right hook. Such vastly different strategies seemed to suit them.

The men shared little in style but plenty by way of their background. Both were born in the eastern United States in the Romantic period, an era in which science was no longer a hobby but a worthy profession—provided you were monied and male. This was something else they shared: they were both beneficiaries of other people's patronage. Cope was born into money, whereas Marsh's financial windfall occurred in his twenties, when—in the style of a Dickens novel—his rich uncle swooped in to pay for his education. If it had not been for their financial backers, it's unlikely either man would have been able to fund their paleontology pursuits. They were in the right place, at the right time, with the right privileges. For a moment, the world was their dig bed.

Being young men with money to burn, in the early 1860s, Cope and Marsh traveled overseas, independently studying specimens of all stripes in museums throughout Europe. Both loved to learn, though their educational journeys took vastly different routes. While Cope's formal education

FIG. 4. Rival paleontologists Othniel Charles Marsh (*left*) and Edward Drinker Cope (*right*), key figures in the Bone Wars. Wikimedia Commons, https://commons.wikimedia.org/wiki/File:Othniel_Charles_Marsh_%26_Edward_Drinker_Cope_bw.jpg.

mostly ended at sixteen, Marsh didn't even begin his full-time schooling until he was twenty-one. At nineteen, Cope had already published in the *Proceedings of the Academy of Natural Sciences of Philadelphia*; at that age, Marsh had barely even left his father's farm.

Yet fate interceded in the winter of 1863, when twenty-three-year-old Cope and thirty-two-year-old Marsh met for the first time in Berlin. Their meeting was cordial, and the pair struck up a professional relationship.

Over the next several years, they exchanged letters, papers, and occasionally fossils. In 1867 Cope's cordiality extended to name a newly discovered amphibian fossil after Marsh: *Ptyonius marshii*. Marsh returned the favor the following year, naming a marine creature after Cope: *Mosasaurus copeanus*.

But Marsh's "honor" came with a twist: the fossil in question had been discovered in a New Jersey quarry to which Cope had introduced him. Unbeknownst to Cope, Marsh had made a deal with the landowner to receive first dibs on future fossil finds, undercutting Cope's claim. Naming one after Cope amounted to throwing his colleague a figurative bone after acquiring all the rest of them himself.

The trespass dumbfounded Cope. He learned quickly—and the hard way—that for Othniel Charles Marsh, fossils were more valuable than friends.

*

While waiting for her mother to pick her up on the last day of school, Ellie spotted a group of friends hop into a parent's minivan and drive off to a birthday party without her.

Cue the quivering lip, the burning cheeks, and the shame of feeling excluded. Meredith pulled into the pickup lane as the van pulled out of sight.

"How was your last day?" Meredith asked.

Sighing, Ellie pressed her forehead to the window.

Of all the human tragedies, not receiving an invitation to a birthday party ranks low on the list. Yet when you're in third grade and your friends are half of your heart, it's hard not to feel devastated. The only thing worse than discovering where you fall on the friendship list is realizing that you didn't make the list at all.

For a week, I wasn't privy to any of this. In my infinite aloofness, all I knew was that her bedroom door seemed closed more than usual. As the days passed, I noticed a newfound sullenness enveloping her. Her eyes lost their shimmer. The happy shrieks from our happy girl had quieted.

It's one thing to lose one friend but quite another to feel like you've lost a minivan full of them. Worse still is never knowing why.

Did she take the last corndog in the lunch line? Nab somebody's preferred library book? When public opinion turns on you, it can be hard to turn it back.

*Thank goodness for the dinosaur trip,* I thought, peering into her room to find her reading on her beanbag chair. *Who needs friends when you've got your dad?*

Even I knew better than to say such words aloud.

*

The beginning of the Bone Wars might be traced to 1869, two years after then-twenty-seven-year-old Cope received a marine reptile specimen from

Kansas. In those days, little was known of what Cope called *Elasmosaurus platyurus*, a relative of the *Plesiosaurus*. Cope reconstructed the bones as best he could with the information available.

In September 1868 Cope discussed the discovery at the Academy of Natural Sciences of Philadelphia and later published his remarks along with an illustration in the society's publication.

Yet, in a career-wounding error, Cope mistakenly placed the skull on the wrong end of the creature's vertebrae while reconstructing the newly discovered species. He'd believed the *Elasmosaurus* had a short neck and a long tail when the opposite was true. Marsh was delighted to point out Cope's mistake, while Cope was loath to believe it. Since neither man budged, they called upon renowned paleontologist Joseph Leidy, perhaps the only scientist in America qualified to determine whether Cope had positioned the creature's head on its hindquarters.

For Cope, the stakes could not have been higher. The up-and-coming paleontologist had tied his reputation to his recently published article. If he were wrong, the entire scientific community would know.

Upon viewing the reconstruction in the Academy of Natural Sciences of Philadelphia, Leidy quickly made his determination. Without malice, Leidy removed the skull and placed it on the opposite end, settling the matter in Marsh's favor.

In an ironic twist, writer Daniel Rains Wallace reports that Cope based his reconstruction on an earlier (and erroneous) sketch of the *Elasmosaurus* by Leidy.

Regardless of the cause of the error, had Cope been capable of rousing some semblance of humility, his career might've continued unabated. Instead, he attempted what Marsh considered a cover-up, buying back every issue of *Transactions of the American Philosophical Society* that featured his article and illustration. Cope's eventual milquetoast attempt at a mea culpa proved wholly insufficient for Marsh, who would never let the scientific community forget Cope's error.

Scientific integrity had been momentarily restored.

But the Bone Wars were just beginning.

*

Outside the windows, the miles pass like seasons. The landscape churns slowly, revealing a pastiche of patterns consisting of grasslands, highways, and clouds.

By 9:30 a.m., we'd left Minneapolis in our rearview. An hour later, we'd passed St. Cloud. By lunch, we'd watched Fergus Falls fade into the distance; shortly after, we bid farewell to Fargo. At a rest area, I make the mistake of glancing at a map to realize that we're still a long way from Bismarck, which—though it sounds like the title of a country and western song—is, in fact, our current predicament.

Given the long road ahead of us, I decide that for the rest of the day, we'll stop for nothing but gas and bathroom breaks—no stretches, scenic selfies, or kitschy roadside attractions.

But then, outside of Jamestown, North Dakota, we pass a billboard that prompts me to reconsider my policy: World's Largest Buffalo Monument.

*Well, maybe just one kitschy roadside attraction . . .*

We follow an understated sign off the highway, eventually winding along the James River on a stretch promisingly named Buffalo Scenic Road. Yet the longer we drive, the less promising our route becomes. The paved road gives way to gravel, which gives way to something not quite road at all. We park along that not-quite road, then peer toward the steep, tree-lined terrain to our left.

"Does that look like a trail?" I ask hopefully.

"Does what look like a trail?" Ellie asks.

We trudge beneath the power lines, then engage in a little light mountain climbing until at last we see it, or almost see it—a smidgen of the sixty-ton, twenty-six-foot-tall concrete buffalo sculpture corralled behind a barbed wire fence.

"We found it!" I say, fully aware that we haven't—at least not by the preferred route. On the far side of the fence, I can make out a few fellow travelers who, having taken the proper path, are rewarded with an unobstructed view of the World's Largest Buffalo Monument.

"It's a good thing it's so big," Ellie sighs, pushing aside branches on the far side of the fence, "or we might've missed it."

Though this seems an inauspicious start to our trip, I hide it well. Ellie and I pose for a selfie, which features a smiling me, a bewildered her, and a sliver of buffalo beard hidden behind a backdrop of trees.

"Perfect," I lie, eyeing the disappointing result.

Ellie shrugs.

"Come on, time to go," I say, quoting my favorite almost country and western song: "We're a long way from Bismarck."

*

For much of the late nineteenth century, Cope and Marsh regularly bickered, battled, and engaged in scientific sabotage in an attempt to ruin one another's careers. Their fights occurred in fits and starts with no clear end in sight. There was always some fossil worth feuding over or some fact needing public dispute. The warring paleontologists had dug a hole of their own making, and they just kept digging.

Most of their battles took place in the West: in Colorado, Wyoming, Nebraska, and Montana—and, in the latter instance, not far from where we'll be. In the mid-1870s, Cope set his sights on central Montana's Judith River badlands, while Marsh and his team oversaw their own expeditions not terribly far away. When their paths crossed, trouble followed. The pair fought over who found what first and what to call it. They fought over dig sites and occasionally resorted to dispatching spies. They hired one another's men, and those in their employ resorted to rock-throwing in at least one instance. Their rivalry worsened; author David Rains Wallace writes that their fierce competitiveness became so great that "dinosaurs came to seem almost beside the point."

Their last battle took place during one of the most stable periods of Cope's life. By 1889 Cope had obtained a professorship in geology and mineralogy at the University of Pennsylvania—a position that offered pay, prestige, and enthusiastic students. However, none of it was enough.

According to paleontologist and Cope biographer Henry Fairfield Osborn, Cope's last battle with Marsh was precipitated by two factors: the secretary of the interior's order requiring Cope to relinquish all fossils found during a government-sponsored survey to the U.S. National Museum, and his encounters with a young reporter named Hosea Ballou, who knew that public feuds sold newspapers. Believing that the secretary of the interior's order to relinquish the fossils was linked to Marsh, Cope let loose the dogs of war.

His weapon: decades' worth of Marsh's alleged mistakes, which Cope had meticulously (and maniacally) compiled and stored in the lower drawer of his study table. With one hand, Cope had discovered dinosaur fossils the world had never seen, and with the other, he'd curated a career-destroying dossier aimed at one of the only other men of his era who dared to do the same.

Cope gave Ballou plenty of dirt on Marsh, and in exchange, Ballou transformed his yellow journalism into gold, publishing Cope's accusations in the *New York Herald*. The various charges leveled against Marsh are too petty and tedious to recount. Nevertheless, Marsh took the bait. The following week, Marsh joined Cope in the mud pit. Marsh's hard right hook came by way of reminding the world of Cope's *Elasmosaurus* error and his own role in setting the record straight.

Cope's "anger was great, and he asserted in strong language that he had studied the animal for many months and ought at least to know one end from the other," Marsh wrote. "It seemed he did not."

Marsh might've been crueler. (After all, Cope didn't know the creature's head from its ass.) Instead, he appeared to have taken the slightly higher road. Either that or Marsh had been fine-tuning his more measured response for some time—as some believed to be the case.

If the latter were true, then the men must have lain in wait for each other for years—one compiling grievances in a drawer, while the other drafted his rebuttal. In hindsight, I can't help but wonder what might have come from their cooperation. What discoveries might have been made together had they chosen a more collegial path?

The complexity of their lives became simpler upon their deaths: two gifted paleontologists whose professional achievements were overshadowed by their grievances against one another. Worst of all, they had dragged much of the paleontology community along with them.

"During the two closing decades of the last century," Henry Fairfield Osborn lamented, "nearly all the vertebrate paleontologists in the United States became partisans in this warfare." Following their embarrassingly public feud in the *New York Herald*, paleontologist and Marsh biographer Charles Schuchert reported that in 1892, the U.S. Senate cut the Geological Survey's budget by 30 percent, including a devastating 50 percent cut to paleontology.

The privilege that Cope and Marsh enjoyed throughout their lives—the chance to dig and discover, often at someone else's expense—became more challenging for others after these dramatic budget cuts. These were men whose contributions to paleontology were unmatched, as was the damage they wrought on their field.

Add further failings into the mix—including both men's racist views on evolution, as well as their tendency to acquire fossils on Indigenous land and their occasional grave robbing—and, by today's standards, they would hardly have received the acclaim they once enjoyed.

It seems a little too easy to call them "products of their time," though they were. Yet they were also men whose personal pettiness seems a bit baffling given their renowned reputations. By professional measures, they appeared to have it all—except for peaceful coexistence with one another.

Among Cope and Marsh's many similarities, I've overlooked one. Both men lost their mothers in early childhood, leaving them primarily in the care of their demanding fathers, each of whom envisioned a very different future for his son.

I will leave the psychoanalysis to Freud; suffice it to say that Cope and Marsh had something to prove. Nothing could sate their need for validation—not a professorship, publication, or dinosaur. Sadly, it's possible that nothing gave them more pleasure than the other's pain.

As a father, I know there's some parental lesson to learn here—something about unconditional love or supporting your children's dreams.

Or maybe something about prioritizing people over one's passion.

While Cope was convinced that his legacy was tied to naming dinosaurs, a greater legacy could have been tied to his daughter, Julia.

The daughter he was separated from for far too long during all those months in the field.

*

The dark comes soon but softly as we put down tent stakes at the Bismarck KOA.

"Dad?" Ellie says, seated atop our site's picnic table.

"Yeah, hon?"

"Can you read *Ramona*?"

I reach for my childhood copy of Beverly Cleary's *Ramona the Brave*, flipping through the front matter until I reach the book's opening line.

Ramona Quimby has served as Ellie's port in the storm in recent weeks. Something about the six-year-old's vulnerability resonates with nine-year-old Ellie. It doesn't hurt that Ramona knows a thing or two about being a younger sibling and trying to be brave when the world feels scarier—and lonelier—than it should.

I like to think of Ellie, our middle child, as "the insulated one." She's dodged the overprotectiveness endured by our eldest and the being doted upon foisted on our youngest. She has one sibling to lift her up and another to cushion her fall. From my perspective, she's landed in the ideal birth order position.

Yet, from her perspective, she's in the worst position—as worshipped neither as the first child nor the last. Instead, she's all jammed up in the middle, the recipient of secondhand everything.

In the old days, when her family got her down, Ellie could always take refuge in her friends. Not anymore.

Like it or not, Dad and Ramona Quimby will likely be her primary companions for the next two weeks. I like to think we are both committed to doing our best.

I'm two chapters into *Ramona the Brave* when I notice my droopy-eyed daughter beside me.

"Hey," I say, dog-earing the page, "how about we brush our teeth?"

"Okay," she yawns.

I lead us on a moonlit walk to the bathroom, and upon our return to the tent, we cocoon ourselves in our sleeping bags.

"Highs and lows?" I ask, beginning our daily tradition.

"Um . . . the high was probably the World's Largest Buffalo Monument we almost saw."

"The World's Largest Buffalo Monument that we *did* see," I correct.

"And the low . . ." she says. "Well, the low was probably leaving."

"Leaving the World's Largest Buffalo Monument?" I ask, straight-faced.

"Leaving our family!" Ellie shouts, finding her smile.

We settle in, surrounded by the silence just beyond the tent.

"Dad?" she asks, burrowing deeper into her sleeping bag.

"Yeah, hon?"

"Can we read one more chapter of *Ramona*?"

"Absolutely," I say, reaching for the flashlight.

Who can say how much time we've got before the next asteroid comes hurtling our way?

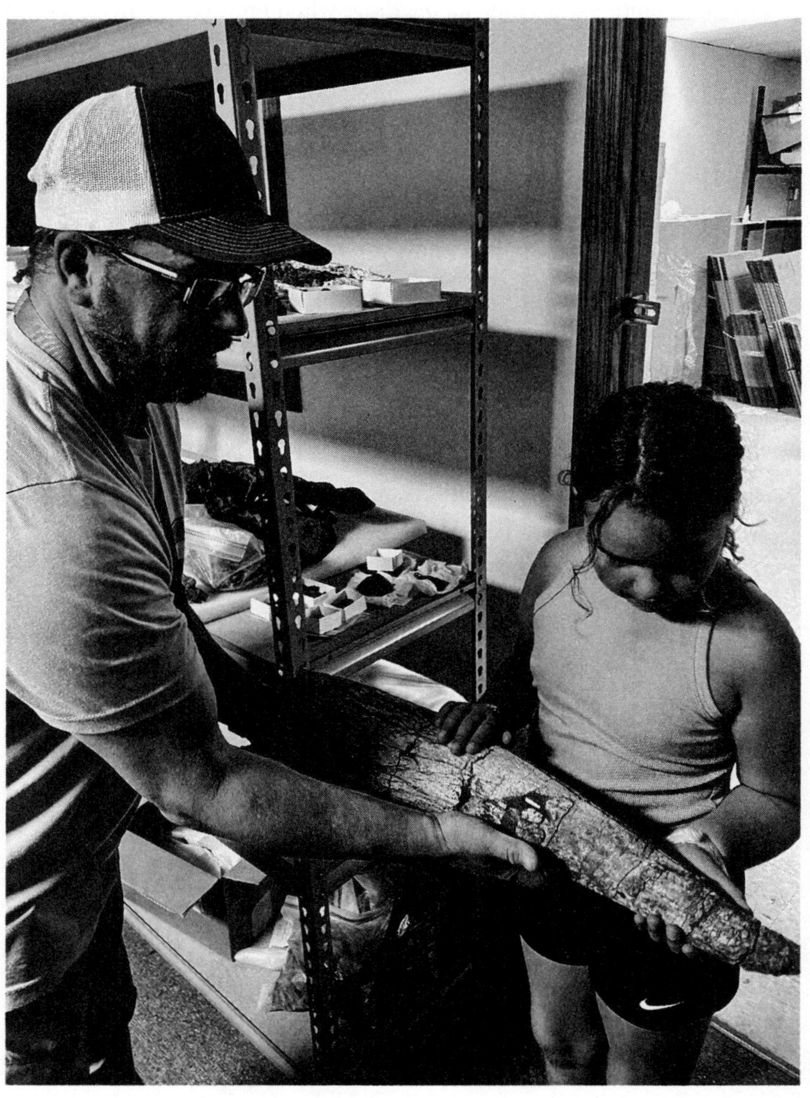

FIG. 5. Tom Hebert shares a *Triceratops* horn with Ellie in the Standing Rock Institute of Natural History. Author photo.

# 2
# Embrace the Evolution

### JUNE 30

*Bismarck,* ND → *Fort Yates,* ND → *Undisclosed Dig Site,* SD → *Billings,* MT

In September 2002, twenty-six-year-old Tom Hebert stumbled into his home in Eau Claire, Wisconsin—drunk, demoralized, and moments away from ending his life.

"From the outsider's point of view, life was good," Tom said. "I was making a hundred grand a year, owned my own house and car, and had a girlfriend. Everything was great," he continued, "except I was drinking every day from sunup until sundown."

Tom reached for his handgun and then sat at his kitchen table.

That's when he heard a disembodied voice—for the first and only time in his life.

"It's not your time," it said.

Tom paused, breathed, and listened to it. He returned the gun to his dresser drawer.

Then he checked into rehab.

Nearly two decades later, in the spring of 2021, Tom enrolled in my rhetoric and composition course—a required course for graduation. I first met him via email—or rather, a barrage of emails—each asking for more information about the class and the assignments. Typically, I wouldn't have minded such curiosity from an enthusiastic student, but since our increasing correspondence occurred several weeks before the start of the term , I found his enthusiasm a bit baffling.

I replied as kindly as I could, reminding him we had plenty of time and that all his questions would be answered on the first day of class.

"I know I am a pain," Tom replied, "but being married with three teenage kids in high school and working full time, I try to get ahead where I can."

A few weeks later, I met Tom in person—a forty-four-year-old in jeans and a pullover and wearing a baseball cap. He sat in the last seat of the last row on the left side of the room. He was the first to arrive and the last to leave. One day, amid our end-of-class banter, he mentioned that he had a fifty-mile drive ahead of him.

"You drove fifty miles?" I asked. "To come to class?"

"It's no big deal," Tom said. "I drive out to Montana all the time."

"You got family there?"

"Nah," he said, revealing an impish grin. "That's where the dinosaurs are."

Tom's passion for dinosaurs began in the summer of 2012, when, in the aftermath of a divorce, Tom turned his attention to his daughters. He wanted to plan a special outing for each, and he wanted them to choose what to do. His seventeen-year-old opted for a country music festival, which Tom obliged, complete with boots and a ten-gallon hat. But it was his eight-year-old daughter's choice that rerouted Tom's life trajectory toward his eventual dream.

"I want to find dinosaurs," she'd said.

They drove west—admiring Mount Rushmore and the Black Hills—before arriving at a commercial dinosaur dig site in South Dakota. By the end of the first day, his daughter had unearthed a *T. rex* tooth.

The father-daughter dinosaur digs became an annual tradition, with more time dedicated to the pursuit each year. Days turned into weeks until, in the summer of 2019, Tom said to his fiancée, Beth, "You know, I think I want to go back to school. I want to put myself in a position to find dinosaurs full-time."

Since Beth was the level-headed one, Tom expected her to pump the brakes.

Instead, she pushed down on the accelerator. "What took you so long?" she asked.

Within months, Tom sold his insurance agency, which had supported him for years, and enrolled in our university.

All of which led him to my rhetoric and composition classroom.

And two years later, it's what leads Ellie and me to Fort Yates, North Dakota.

*

Shortly after sunrise, Ellie and I jam our sleeping bags into stuff sacks, refold tent poles, and buckle ourselves into the van. We are less bright-eyed and bushy-tailed than bleary-eyed and beat.

"Why is it so bright out?" Ellie moans, adjusting her seatbelt. "What's wrong with the sun?"

"It's just doing its job," I yawn.

There's no coffee cup big enough for the day ahead, which includes 450 miles of driving, interspersed with interviews and dinosaur digs.

Our first stop: the Standing Rock Sioux Tribe's Prairie Knights Casino and Resort, where Tom Hebert had stayed the previous night.

The enormous structure—featuring seven hundred gaming options, from slot machines to poker tables—contrasts with the silence of the open plains. However, given that it's barely 8:00 a.m., even the loosest slot machines remain silent.

"Well, look who it is," Tom calls, strolling steadily down the corridor. He's wearing his Earth Sciences Foundation trucker hat and matching ESF shirt. Since 2021 Tom has served as the founder and director of Earth Sciences Foundation Inc., a nonprofit whose mission is to "make the world of Earth Sciences available to everyone."

"Tom," I say, extending my hand. "We're a long way from the composition classroom, huh?"

"We sure are," Tom laughs, the casino lights blazing behind him. He turns toward Ellie. "Hey, Ellie, you want to see something cool?"

"Sure!"

"Come on, they've got some dinosaur bones over here."

The Standing Rock Sioux Tribe owns and operates the casino and resort, which features various culturally significant artifacts, from artwork to dinosaur fossils. We admire the latter from behind the glass—a collection of fist-sized vertebrae fossils, mainly from duckbilled *Hadrosaurs*.

The Standing Rock Sioux's land has long been a hotbed for fossil activity, most notably the Hell Creek Formation, which—during the Late Cretaceous—resembled a braided-stream-filled swampland well suited for an array of dinosaur species to thrive. For decades, the land has been home to a treasure trove of newly discovered dinosaur fossils.

While Ellie and I are in awe of the fossils, Tom is less so.

"If you like this," Tom chuckles, "just wait till we get to where we're headed next."

Fifteen miles later, we arrive at Fort Yates—the tribal headquarters for the Standing Rock Sioux Tribe. Over the past two years, Tom has become a regular visitor to Fort Yates, a hamlet within the tribe's 2.3 million acres between North and South Dakota. With a population of around 175, Fort Yates is no booming metropolis, though as the tribal headquarters—not to mention the county seat—it is a hub of power.

Fort Yates is also the burial site of Tatanka Iyotake (Sitting Bull), the Hunkpapa Lakota leader whose legacy is widely remembered for his fearlessness and commitment to preserving tribal culture by avoiding reservations—much to the frustration of the U.S. Bureau of Indian Affairs.

Our first stop in Fort Yates is to pay our respects to Tatanka Iyotake. A small sign directs us to a placard leaning against a boulder on the town's main drag. We exit our vehicles and stand before the placard, which recounts Tatanka Iyotake's vision of defeating white soldiers—a vision fulfilled in 1876 at the Battle of Greasy Grass (also known as the Battle of Little Big Horn).

"If you want," Tom says, "I can teach you a little Lakota."

We turn to attention.

"Hihanni waste," Tom says. "It means 'good morning.' And if we turn to the east," he continues, positioning himself toward the sun, "we can say good morning to Unci Maka, Mother Earth."

"Where'd you learn that?" I ask.

"I'm fortunate to have some Standing Rock Sioux friends willing to teach me," Tom says. "Plus, they gave me a Lakota dictionary."

Tom's relationship with the Standing Rock Sioux Tribe extends beyond the occasional language lesson. In 2022 Tom's newly formed nonprofit partnered with the tribe, granting Earth Sciences Foundation exclusive access to dig for dinosaur fossils on tribal lands and help document and preserve all dinosaur fossils displayed in Fort Yates's now-shuttered Standing Rock Institute of Natural History. Tom believes that the Standing Rock Sioux's dinosaur fossil collection is one of the premier collections in the region. However, without a significant change, few will ever see it.

"So what does the partnership mean?" I asked Tom shortly after the announcement was made public.

"Well, for one thing," he said, "I'm one of the only white people the tribe trusts to do the work. I just hope I can do it well enough."

That a white guy from Wisconsin (with no prior experience with tribal communities) is now the primary paleontological partner with the Standing Rock Sioux seemed rather unexpected.

"Tom," I said, treading lightly, "how did you persuade the tribal council to trust you?"

"Well, Professor Hollars," he said (insisting on the title), "it's just like we always talked about in class: know your audience."

"What'd you say?"

"I told them that white people like me had been screwing them over for generations. And it was high time we tried to make it right."

After much discussion, the tribal council voted to approve a five-year partnership with Earth Sciences Foundation. That partnership allows Tom, Ellie, and me to continue the dinosaur portion of the Fort Yates tour at the once-thriving Standing Rock Institute of Natural History, which has remained closed on the outskirts of town since 2019.

Back in the vehicles, we drive one mile west from Tatanka Iyotake's burial site to the Standing Rock Institute of Natural History.

"Right this way," Tom says, unlocking the gate and allowing us entrance into the institute's gravel parking lot.

As we exit the van, Ellie and I peer out at the institute—its entryway engulfed by tufts of hip-high grass. Tom leads us toward a side entrance, unlocks the door, and enters the darkened fossil repository.

In 2007 the Standing Rock Sioux's paleontological future had looked bright. That was the year tribal member Adrienne Swallow founded the Standing Rock Sioux Tribe's Paleontology Department, partnering with Sitting Bull College to take students to dig sites throughout tribal lands. Those digs yielded more than ten thousand fossils representing a wide range of Mesozoic-era animals, from *Tyrannosaurus* and *Triceratops* to ancient crocodiles and fish. And those initial discoveries barely scratched the surface.

While tribal members were aware of the lucrative natural resources buried beneath their land (including lignite, natural gas, and geothermal energy), Adrienne Swallow turned her attention to a different readily available resource: dinosaur fossils. In 2011 the tribal council adopted the first tribal paleontological resource code in the country, formally recognizing the "scientific, economic, recreational and cultural value" of their paleontology resources.

Plans for the Standing Rock Institute of Natural History were well under way when Swallow died of cancer in 2015. Despite efforts to continue Swallow's vision, the museum never achieved its full potential without her.

"I always like to announce my presence before I enter," Tom explains, cracking wide the fossil repository door. "Hello!" he calls. "Is anybody here?"

We're met with silence, so he pulls the door wider, granting us access alongside him.

Our eyes fall to a room full of shadow-laden plaster casts stacked halfway to the ceiling.

"What's in the casts?" I ask, nodding toward them.

"Dinosaur bones," he says. "Mostly *Triceratops* and *Edmontosaurus*, but supposedly, there are a couple of *T. rex* in here, too."

Most of these fossils were collected in the late 1990s and early 2000s by students from Sitting Bull College. Since the institute closed and the college's paleontology program remains in limbo, few people can do the prep work. However, members of the Standing Rock Sioux Tribe leadership team dream of building a new museum to showcase their collection, not only for cultural and scientific purposes but also for economic ones linked to dinosaur tourism.

Tom leads Ellie and me from one repository room to the next until we reach the main exhibition hall. With the electricity turned off in this section of the institute, we navigate the darkness with heart-pounding anticipation, never entirely sure what awaits us around the next bend.

What awaits us, of course, are dinosaurs. First and foremost, a replica skull cast of Stan the *T. rex*. As *T. rex* discoveries go, Stan ranks second only to Sue—the physically largest and most complete *T. rex* ever found. But Stan holds the distinction of having the best-preserved skull, making it the crown jewel in his new home at Abu Dhabi's Department of Culture and Tourism. In 2020 Stan was sold at auction for $31.8 million—nearly four times the price Sue fetched just twenty-three years prior. Which makes Stan, to date, the most expensive fossil ever sold.

Before Stan, Sue, and the era of dinosaur bidding wars, fossils were hardly big business. Rock shops did all right, but dinosaurs rarely made their way to the auction block. That changed in 1997, when Sue was sold to Chicago's Field Museum for $8.36 million. These days, five or so dinosaurs are auctioned annually, and they regularly command price tags in the millions. While dinosaur fossils were once mainly within the purview of paleontologists, now paleontologists must compete with private collectors, many of whom care less about scientific discovery than having a dinosaur in their foyer.

But for those fortunate enough to see dinosaur fossils up close—as we do now—it's easy to overlook such complexities in favor of the pure wonder dinosaurs provide.

"This is the greatest job ever," Tom sighs, gazing at the *T. rex* skull staring back at him. "I go to bed every night thanking God for the blessings I've had."

*

To understand the origins of America's prehistoric past, look back long before Cope and Marsh's Bone Wars. Indigenous people have been finding fossils throughout North America for thousands of years, often weaving their discoveries into their stories and culture.

In *Fossil Legends of the First Americans*, historian and classical folklorist Adrienne Mayor writes that "even though Native American understandings of the fossil record were not scientifically methodical in the modern sense, they offered an alternative, coherent way of interpreting earth's history." The Plains Indians imbued certain marine fossils with magical properties to "summon buffalo herds," writes Mayor, adding that members of the Sioux tribe may have viewed fossil discoveries as proof of "Thunder Birds and Water Monsters."

Though it's difficult to put dates on beginnings, Mayor traces one of the earliest documented discoveries of prehistoric fossils in modern-day America to 1739, when a hunting party of Indigenous warriors (likely of the Abenaki tribe) shared with their French Canadian soldier allies what they'd discovered in the marshy swamps in Big Bone Lick along the Ohio River near modern-day Kentucky's northern border. The soldiers stood, wonderstruck, as the warriors pulled forth from their canoes what Mayor described as "an enormous, fossilized femur nearly as tall as a man, several huge teeth, and great ivory tusks darkened by time."

In 1762—twenty-three years after the discovery—French naturalist Louis Jean-Marie Daubenton presented a paper on the fossils to the French Royal Academy, arguing that they belonged to two living species: the femur and tusk were of the elephantine family, and the molars from a carnivorous hippopotamus. Although the location where they were found had no history of such creatures, his guess was as good as anyone's.

After all, at the time of the fossils' discovery, neither dinosaurs nor the concept of extinction was known. What seemed more logical? That the fossils belonged to fauna *known* to this world, or that they were proof of ancient creatures the likes of which few had ever dreamed? Daubenton was right to note the fossils' elephantine qualities. However, he could not follow his lead to its correct conclusion—that they belonged to the American mastodon, the first complete skeleton of which wouldn't be discovered for another eighty-three years. Simply put, we had the fossils before we had the language for the creatures from which they'd come.

It's not uncommon for scientific discoveries to transcend lifetimes.

Or for our understanding of the world to evolve.

*

At a few minutes past 10:00 a.m., Kerry Libby—the tribe's external affairs director—meets us near the entrance of the shuttered exhibition hall.

"Welcome," Kerry says, turning her attention immediately to my daughter. "You must be Ellie."

"Yup!" Ellie says.

"So what do you think of our dinosaurs, Ellie?" Kerry asks.

"Really cool," Ellie says.

"You know," Kerry smiles, leaning close, "I think so, too."

Ellie and I trail Kerry back into the darkened institute.

"My whole life, I've loved dinosaurs and digging up rocks. I've always loved having my hands on the earth and being a nerdy girl. And there's nothing wrong with being a nerdy girl," Kerry says, turning to Ellie. "Got it?"

Ellie nods.

"I liked to do these things but never got involved in paleontology or geology," Kerry continues. "Not until Adrienne Swallow came into my life."

In 2014, when Swallow informed Kerry they needed someone to serve as the institute's first director, Kerry reminded her that she wasn't a paleontologist.

"You don't have to be," Swallow assured. "We need you to run the operations."

Having recently completed her master's degree in strategic leadership, Kerry agreed to take on the directorship. But eight months after Swallow died in 2015, Kerry was diagnosed with pancreatic cancer.

"I was forced to resign," Kerry says.

For a time, the Standing Rock Sioux's dreams of a robust paleontology program seemed to be on hold. However, when Tom Hebert approached the tribal council in 2021 about a potential partnership, members felt hopeful yet cautious. Could this white guy from Wisconsin revitalize their program?

While some council members were supportive, others were opposed—not to Tom's involvement, but to unearthing ancient fossils under any circumstances. They argued that disrupting the dinosaurs' burial sites would risk releasing evil spirits into the world.

Unsure of how to proceed, Kerry sought guidance from the tribe's medicine people.

The dinosaurs "want their story told when they come out of the earth on their own," one of the medicine people assured her. "You're not digging them up on purpose; they're exposing themselves to you."

Kerry listened to the medicine person and concluded that they weren't just digging up dinosaur fossils but also unearthing scientific and cultural stories.

"All of these things have a spirit with them," Kerry says, extending her hands toward the collection's many fossils. "They're still connected; that's what our culture really believes. When something dies, it doesn't just die. So you have to do the things you have to do to show them respect."

During Kerry's tenure as director, she and her team conducted a weekly smudging ceremony. They would walk from one corner of the institute to the other, burning sage to purify the space.

"We begin by acknowledging Unci Maka—Mother Earth," Kerry explains. "But we have to acknowledge everything else, too—the birds, deer, elk, and all the things down to the turtles and snakes that crawl. We say a prayer before we travel to protect the winged, the two-legged, the four-legged, and the crawling animals. And we have to give that same sort of respect to these things that existed millions of years ago."

Tom had previously shared that throughout the tribe's discussions on dinosaur fossil extraction, he'd tread lightly.

"I told them that I'm not here to tell them their beliefs, their traditions; that's none of my business. But my thought was, 'If you're concerned about evil spirits getting out, you have two options. Mother Nature will root them out in an uncontrolled manner, or we can try to control it and mitigate the risk.'"

As Tom busies himself with a display, I turn toward Kerry and ask, "Of all the partners you might have worked with, what made you choose Tom?"

"One word," Kerry smiles. "Persistence."

Kerry likes Tom's hustle. And his steadfast commitment to the tribe.

Additionally, unlike many professional paleontologists—who can demand hefty fees—Tom, an amateur paleontologist, works for the tribe for free.

"He said, 'I won't charge you anything,'" Kerry says. "'I just want to help.' And that's what's happened."

Tom's agreement with the tribe is clear: Earth Sciences Foundation has exclusive access to dig on Standing Rock Sioux tribal lands under the condition that whatever Tom and the foundation find stays with the tribe.

What does Tom get out of the partnership? The chance to discover dinosaur fossils in some of the richest fossil beds in the world.

What do the Standing Rock Sioux get? Tom and his foundation's help in securing fossils and managing their vast collection. Which, Kerry knows, has the potential to pay enormous dividends in terms of the region's dinosaur tourism industry.

"We get thousands of people who come to Fort Yates just to see where Sitting Bull is buried," Kerry says. "But then you've got this whole other demographic of young people—like little young Kerry Libbys and Ellie here—who can help dig dinosaurs."

Kerry dreams of weeklong digs on tribal lands, with campfire dinners and nights spent in tipis. A culturally sensitive, authentic experience led by the tribal members themselves.

"If you have the right person in the right position," Kerry says, "it could be a huge revenue stream."

Tom had made much the same case to me shortly before Kerry's arrival. "If we could get a hundred thousand people out here, multiplied by thirty dollars per ticket, that's three million in admission alone," he'd told me as we roamed the darkened institute. "Think how much that could help the tribe."

Turning to Kerry, I say, "It seems like you might be the right person."

"Maybe someday," she says. Though for now, Kerry's role as the tribe's external affairs director keeps her plenty busy with more pressing matters, from public safety to health and wellness.

"That's why I rely so heavily on Tom," Kerry says, nodding toward him on the far side of the room, straightening a placard. "He sees my vision, and he understands it."

Together, we step outside as Tom locks the institute's doors behind us.

"What's next on your agenda?" Kerry asks.

"We've got some digging to do," Tom says.

"Well," Kerry says, giving Ellie a high five, "go out there and find a big one."

Waving goodbye to Kerry, Ellie and I settle into our van as Tom starts his truck.

"Ready to see my office?" Tom shouts out the truck window.

"I think so," I shout back.

"No offense, Professor Hollars," Tom says, "but the view from *my* office is a teensy bit better than the view from yours. But don't worry," he laughs as he pulls away, "because today, it's your office too!"

*

After a short drive, we enter South Dakota, parking our two-vehicle caravan miles from the nearest major road.

"Here we are," Tom says, though I have no idea where "here" is.

We peer at outcroppings, buttes, and tufts of weeds—a rugged paradise as far as the eyes can see.

Tom's wearing his field pack, into which he's crammed several dig kits, a yellow waterproof field notebook, a couple of pencils, a roll of aluminum foil, and—to firm up any splintering fossils—a tube of commercial-grade cyanoacrylate adhesive.

"Searching for fossils is like piecing together the world's worst puzzle," Tom says as we tromp through the dust. "No edge pieces, no box, no nothing."

Mere minutes pass before the puzzle pieces begin to reveal themselves. Tom points to our left and right toward various fossil fragments, which are everywhere.

"That's a chunk from a sixty-six-million-year-old tree," he says, "and that white stuff over there, that's a piece of exploded dinosaur bone."

For a guy like me, who's never seen a dinosaur fossil in the wild, it's all a bit overwhelming. Yet Ellie takes it all in stride, crouching alongside a white, chalky fragment within minutes.

"What about this?" Ellie asks, picking up the fragment. "Is this dinosaur?"

"I don't know," Tom asks. "Give it a lick."

Ellie eyes me. Should she really lick a potential dinosaur fossil she'd just picked up from the ground? Though it doesn't pass the five-second rule (or even the fifty-million-year rule), Ellie extends her tongue and licks.

"Sticky," Ellie confirms.

"That means it's a fossil," Tom says. "Fossils are porous, so when you lick one, it'll stick to your tongue."

Ellie laughs. Maybe finding dinosaur fossils isn't so difficult after all.

"But this one," Tom says, taking a closer look, "isn't technically a dinosaur. It's from an ancient croc that lived around the same time, Late Cretaceous."

"How can you possibly know that?" I ask.

"You see how the bone makes a little X here?"

I nod.

"That means it's a vertebra from a *Champsosaurus*," Tom explains.

(A species, I'll learn, first discovered by Edward Drinker Cope during his 1876 expedition to Montana).

"So, what exactly are we looking for?" I ask Tom, eyeing a sedimentary formation.

"White, chunky bone fragments," Tom says. "When you find those, you look up toward the nearest outcrop and start digging."

The trick to finding dinosaurs, Tom says, is understanding the geology. If the rock is less than sixty-six million years old, then you aren't going to find any dinosaurs. But if the rock is sixty-six million years or older—and you're in the right locale—there's a chance you might. The good news, Tom explains, is that you don't need fancy equipment to read the age of a rock.

"All you need," Tom says, "is the K-T boundary."

Also known as the Cretaceous-Tertiary boundary (or the Cretaceous-Paleogene boundary), the K-T boundary is the convoluted and ever-evolving name for the geological signature embedded within rock layers that marked the end of the Cretaceous period and the beginning of the Paleogene period. When the asteroid (formally known as the Chicxulub impactor) crashed into Earth, it left behind more than a cataclysmic disaster. It also deposited high levels of the chemical element iridium, visible today through a dark band of rock separating layers of sediment. Although the boundary measures only a few centimeters thick, it grows in visibility when paired with dark sedimentation. As a result, if you see the K-T boundary and dig beneath it, you know you've hit potential dinosaur pay dirt. Any fossils found above it aren't from dinosaurs.

Five minutes into exploring the terrain, Tom crouches to study what he believes to be a significant amount of crumbling dinosaur "float"—bone fragments from larger fossils. Its sun-bleached white coloration (rather than dirt-clod brown) is the first clue to what we're dealing with, and a lick test confirms it's a fossil.

"Based on the float and the geology," Tom says, "my guess is we've got a potential dinosaur bone somewhere on that shelf there."

FIG. 6. Ellie digs for a *Hadrosaur* fossil in an undisclosed location in South Dakota. Author photo.

Tom unzips his pack and hands us our homemade dig kits—plastic pencil boxes into which he's placed a scalpel, screwdriver, pick, and paintbrush.

"These are perfect for taking kids out on digs," Tom explains. "Right, Ellie?"

"Uh-huh," she says, her eyes laser-locked on the potential fossil in the shelf before us.

For Ellie and me, this is serious business. We are just one *Tyrannosaurus rex* away from finding a *Tyrannosaurus rex*.

"This looks kind of like a dinosaur bone," Ellie says, crouching over an inch-by-inch square embedded into the rock.

"That's because it is," Tom says, lending a few flicks of his paintbrush to the effort. "Look at you," he laughs, "you're a dinosaur magnet!"

Beaming, Ellie reaches for her pick.

"That's good," Tom says, "the closer to the bone, the smaller the tool. Just take what the rock gives you. And take your time. I always tell people, 'That bone's been there for sixty-five million years. It's not going anywhere.'"

Ellie switches between pick and paintbrush, gaining more confidence every second. After fifteen minutes, she exposes three inches of dinosaur fossil, with more on the way.

Spotting the splintering bone fragments, Tom reaches for a tube of cyanoacrylate adhesive, which he generously coats over the fossil's exterior.

"When you first find a bone, you get all excited and think, 'I hope it's something big!'" Tom chuckles. "But once the sun starts going down, and you're losing light, you think, 'I hope it's something small! I need to get this out of the ground!'"

For now, however, we have a more pressing problem than daylight. In the distance, a rumble of thunder all but shakes the earth.

"We want to get out of here before the rain hits," Tom says, eyeing the darkening sky. "The road gets sloppy out here because of the bentonite clay. You're pretty much stuck if that stuff dries in your tires."

Since we prefer scenarios that do not involve us becoming stranded miles from the nearest major road, we pick up the pace—three picks and paintbrushes working fast to reveal eight inches of what Tom believes is a piece of a *Hadrosaur* rib.

"Now I open my handy-dandy app," Tom says, tapping his phone and placing it directly alongside the fossil, "which gives me the exact coordinates, elevation, and orientation."

Once the data is digitally secure, he snaps a few photos and uploads them to the app.

"Now we're going to tinfoil that sucker," Tom says, tearing a metallic sheet from the roll. "I call this a poor man's plaster." Tom digs a tiny trench beneath the bone before removing it and wrapping it in foil.

"There," Tom smiles, handing Ellie a foil-wrapped fossil. "You're a rock star. I'm so proud of you."

Ellie's speechless.

"You just found your first dinosaur bone!" Tom says. "Do you have any friends back home who can say that?"

Bashful, she shakes her head no.

She stares at the bone, marveling at what she has unearthed. For the first time in a long time, she's not "just" the middle child. She's got a new title: discoverer of dinosaur bones.

"And no one can take this experience from you," Tom says, looking her in the eye. "Not anyone."

"So what happens to it now?" I ask.

"Now I clean it up and place it in the Standing Rock Institute of Natural History," Tom says. "Heck, maybe one day it'll have a little card beside it that says, 'Discovered by Eleanor Hollars.'"

Ellie beams.

Overhead, one more rumble of thunder as darkness descends.

"Time to roll," Tom says, gently placing the fossil in his backpack.

He doesn't have to tell us twice.

*

In the sliver of the rearview mirror, the sky roils from gray to black. Slowly, the storm gains on us, closing in on three sides of the van. We can't drive fast enough. We are surrounded. The clouds reach toward the ground like cupped hands. Aside from the occasional barn or silo, telephone poles are the only signs of human life.

Yet somehow, we are spared the deluge. A drizzle coats the van and Tom's truck just ahead of us, but not so much that the drops can't be flicked clear with a swipe of the wiper blades.

Ellie stares out the window, watching the country unfold from prairies to foothills. We've been driving for hours and still have a few more to go before Billings. She yawns. We're both in desperate need of a diversion.

"Want to play the Alphabet Game?" I ask.

Ellie perks up. She loves scanning road signs and billboards in search of all the letters in their proper order.

Yet out here, there are no signs, billboards, or letters. One glimpse of the landscape confirms it.

"Da-ad!" she groans as the truth sets in. "We can't play! Not out here."

"We could," I laugh. "It just might take a while."

We arrive in Montana at 4:06 p.m. It's only day two, but already, we've nearly logged a thousand miles.

We follow Tom to a rest area near Miles City, where we pause to peer out at the Yellowstone River below.

"That river will follow you all the way into Billings," Tom tells us. "I'll peel off toward Roundup soon, but you guys just stay on I-94 until you hit the city. The river will lead you."

"Got it," I say. "And thanks for a great day, Tom."

"Thank you guys," Tom says. "We'll see you in Roundup tomorrow afternoon."

This is, indeed, the plan: spend the morning visiting the first two stops along the Montana Dinosaur Trail, then connect with Tom in the central Montanan city of Roundup by midafternoon.

Back on the road, Tom peels off as promised, but not before a couple of farewell honks.

FIG. 7. Charles Knight's 1897 watercolor painting *Leaping Laelaps* depicts a more dynamic interpretation of dinosaurs. Wikimedia Commons, https://upload.wikimedia.org/wikipedia/commons/8/8f/Laelaps-Charles_Knight-1897.jpg.

"Hey, Dad?" Ellie says after a few minutes of silence.

"Yeah, hon?"

"Remember when I found that dinosaur bone?"

"Yeah, hon," I smile. "I do."

*

Until the late 1960s, dinosaurs got a bad rap. For over a century, the scientific community had pegged them as slow, dumb, cold-blooded, and lizard-like creatures. One exception came not from a paleontologist but from a wildlife and paleoartist named Charles Knight, who in 1897 famously painted *Leaping Laelaps*, which depicted a pair of dinosaurs (known today as *Dryptosaurus*) engaged in frenzied combat—one frozen in midair while leaping toward its rival. Knight's rendering offered the most dynamic depiction of dinosaurs to date, one that paleontologist John Ostrom would champion more than half a century later.

In 1964, while traversing Montana's Bighorn Basin near the town of Bridger—roughly forty-five miles south of our destination in Billings—Ostrom and his team discovered 124-million-year-old fossilized remains, including a claw of a dinosaur that would become known as *Deinonychus*, the inspiration for *Jurassic Park*'s *Velociraptor*. Upon studying *Deinonychus*'s posture and hunting habits, Ostrom hypothesized that the creature was more suited as a warm-blooded rather than cold-blooded animal.

While early twentieth-century paleontologists assumed the dynamic dinosaurs showcased in Knight's *Leaping Laelaps* resulted from the artist's creative license, Ostrom's *Deinonychus* discovery laid the scientific groundwork for a paleontological paradigm shift. What if dinosaurs weren't slow, dumb, cold-blooded, lizard-like creatures? What if they were smart, warm-blooded, and more closely related to birds than reptiles?

For decades, Ostrom's protégé Robert Bakker continued to advocate for many of his mentor's hypotheses, contributing to our understanding of dinosaur physiology. Additionally, Bakker did much to enhance the flagging public perception of dinosaurs.

"Dinosaurs have a bad public image as symbols of obsolescence and hulking inefficiency," Bakker wrote in *Scientific American* in 1975. He argued that recent research indicated that dinosaurs were "more interesting creatures, better adapted to a wide range of environments and immensely more sophisticated."

Most groundbreaking of all was Bakker's reassertion of Ostrom's theory that dinosaurs—"the epitome of extinctness"—perhaps weren't extinct at all. "The evidence suggests that, in fact, the dinosaurs never died out completely," Bakker wrote. "One group still lives. We call them birds." English biologist Thomas Henry Huxley had first theorized a connection between dinosaurs and birds a century before. But it wasn't until Ostrom, and later Bakker, that the theory gained traction.

Over the next few decades, our previous understanding of lizard-like dinosaurs was replaced by new thinking that at least some subgroups of non-avian dinosaurs exhibited feathery features rather than scales.

Suddenly, dinosaurs were more than dusty fossils in the ground. They were also the birds in our backyards. After decades of decline, by the late 1960s, Ostrom and Bakker's research renewed scientific and public interest in dinosaurs.

An era soon to be known as the Dinosaur Renaissance.

*

We arrive at the Billings KOA an hour before the pool closes.

"Come on, Dad!" Ellie calls from outside the tent. "Hurry up!"

Ellie, part mermaid, cannot go a day without immersing herself in water.

I tighten my swimsuit, then unzip the tent, and sprint after her toward the pool.

We swim until 9:00 p.m., a baptism by chlorine that washes the dust from our skin. Returning to the tent, we change into pajamas. Then, as the sun dips beneath the trees, we take our nightly constitutional around the campground.

I spot a narrow path cutting through the tree line just beyond the RVs.

"Feel like exploring?" I ask.

Ducking through a canopy of branches, we arrive at a small stream over which some previous passersby have laid a plank. We tightrope-walk across it and are rewarded with an unobstructed view of the fast-flowing Yellowstone River. We'd seen it from afar back at the rest area in Miles City, but now we see it up close. Something like wonder crosses Ellie's face as she wades into its ankle-deep water. I feel it too. We're sharing a piece of the world that we have found together. A small, secret place that required a literal walk across a plank.

Then—as the last of the light fades beneath the million-year-old Rimrocks—a change occurs. Just like that, the kid is gone, and a preteen Ellie appears in her place. I am stunned by the transformation, which has undoubtedly been occurring for weeks but is only catching up with me now. Why is it that evolutionary change can take millions of years within species but happens in a heartbeat when it's your daughter?

"What?" Ellie asks, brushing a wayward curl behind her ear.

"What?" I repeat, looking away.

The sun takes its bow on the far side of the embankment, and smooth stones gather in the sand near the shore.

I'm feeling a little woozy, a little fuzzy in my skin, so I pick up a stick and inscribe our names in the sand.

"It's us," I say.

"This is us too," she says, adding her footprints alongside mine.

As Ellie skips stones, I repopulate the world with dinosaurs. I rewind the planet sixty-six million years to this spot, imagining a herd of *Hadrosaurs* foraging through the foliage, their duckbills searching for food. In the distance, several *Triceratops* slowly bed down for the night. Maybe a couple of hardheaded *Pachycephalosaurs* race across the swampland.

I wake from my daydream to find Ellie turned toward me, her face aglow in moonlight.

"Ready?" she asks.

*Never,* I think.

Together, we make our way back toward the tent.

FIG. 8. B.J. and Ellie pose alongside Big Mike outside Museum of the Rockies in Bozeman, Montana. Author photo.

# 3
# Dollars for Dinos

**JULY 1**

*Billings,* MT → *Bozeman,* MT → *Harlowton,* MT → *Roundup,* MT

As Ellie and I climb the steps toward the entrance to Bozeman's Museum of the Rockies, we encounter our first *T. rex* of the day.

I freeze, placing my hand in front of Ellie, stopping her midstep. "Do you think he sees us?" I whisper.

Ellie does not dignify my joke with a response. Instead, she struts up to Big Mike (as he's known to his friends)—and waves at the world's first life-size bronze *T. rex*.

"Hi, Big Mike," she calls. "I'm Ellie."

"Come on," I say, leading us into the lobby. "It's time for the real thing."

Inside, we hand our Montana Dinosaur Trail passport to a woman behind a desk.

The passport—identical in size and shape to a regular international passport—provides another incentive for people like us to complete the MDT. At each trail stop, we'll collect a stamp (featuring the location's "signature dinosaur"), and upon collecting them all, we'll be rewarded with the greatest of gifts—a genuine, one-of-a-kind free MDT T-shirt.

The woman behind the desk flips through the passport's pages until she arrives at the designated space for the Museum of the Rockies, which features a *Torosaurus*.

"First stop," she smiles, pressing the stamp firmly. "You've got a loooong way to go."

Don't we know it.

The MDT's origin story begins with Victor Bjornberg, the Montana Office of Tourism's tourism development and education coordinator. In the early

2000s, he partnered with Montana State University's Extension Service to offer the small town of Malta (three hundred miles north of Bozeman) a program called the Community Tourism Assessment.

"The whole idea was to help communities decide what they wanted to do with tourism, if anything, for their economic development," Victor explained to me. "Malta had been finding some incredible dinosaurs, and they decided, 'Well, we've got these, so let's develop our local tourism based on paleontology.'"

Part of that development involved leveraging the many dinosaur discoveries in central and eastern Montana. Victor recounted that the Malta planning committee acknowledged their community was a "long way from a lot of places," and convincing visitors to drive for hours just to see a few dinosaur fossils would be challenging. However, if they created a "critical mass" of dinosaur attractions and paleo experiences by connecting the existing Montana dinosaur-themed museums into a trail of "paleo gems," their chances for success could improve.

The committee agreed that Bozeman's Museum of the Rockies in the state's southern half was a natural anchor point. Serendipity presented the other anchor point when the U.S. Army Corps of Engineers broke ground on the Fort Peck Interpretive Center four hundred miles to the northeast. Following the 1997 discovery of Peck's Rex (one of the most complete *T. rex* skeletons ever found), federal funding was granted for constructing the interpretive center, a site dedicated not only to the region's dinosaur discoveries but also to the history of the Fort Peck Dam—one of the largest public works projects of the New Deal. Twelve other MDT sites are between Bozeman and Fort Peck, creating a loop of sorts that covers around two-thirds of the state.

The Montana Dinosaur Trail timed its unveiling to coincide with the grand opening of the interpretive center. In May 2005, as ribbons were cut at Fort Peck, the first wave of dinosaur enthusiasts packed their bags and headed for Montana.

The MDT and its array of partners made for odd bedfellows ("I worked at the tourism office for twenty-two years, and I don't remember anything else like it," Victor said, referring to the consistent promotional and funding support from state and federal agencies, tourism regions, and trail facilities), yet dinosaurs have a way of defying expectations.

More surprising than the all-hands-on-deck support is that the Montana Dinosaur Trail is an entirely voluntary partnership.

"This is not a formal organization," Victor explained. "This is a group that functions on a handshake."

It seems impossible in the modern era, yet according to Victor, it works.

"Each facility recognizes the value of working as a group," Victor said. "Your job is not just to promote yourself; it's to promote everyone else too."

The same herd mentality allowed Montana's duckbilled *Maiasaura*—and so many other species—to thrive. When you're not the apex predator, partnerships can help ensure survival.

By any measure, the Montana Dinosaur Trail has been a resounding success for the state, particularly for those towns east of the Rockies. While Montana's western side has little trouble enticing tourists (two words: Yellowstone and Glacier), central and eastern Montana have historically struggled. Yet since the trail's inaugural year, the number of MDT visitors has trended upward. In 2005 a combined 211,634 people visited the fourteen MDT stops; by 2021 that number had nearly doubled, to 409,676.

For a state with a population of 1.1 million residents, attracting 409,000 visitors—many of whom were lured to the less-populated side of the state by dinosaurs—is no small feat. Nor is the visitors' economic impact. A recent economic impact study from the state's Office of Tourism reported that in 2021, 12.5 million nonresidents visited Montana, contributing over $5.1 billion to a state whose combined state and local revenue for that year was $13.1 billion. Meaning that well over a third of all state revenue came from out-of-state tourists—hundreds of thousands of whom flocked to the Treasure State for the Montana Dinosaur Trail.

Dinosaurs and tourism are big business. And when you combine them, you've got an industry.

But like most industries, dinosaur tourism comes with its own brand of competition.

Everyone wants the biggest, most complete dinosaur—but not everyone plays by the same rules.

*

With our MDT passport stamp secured, we make our way to Zach Perry, the museum's assistant store manager and a newly accepted graduate student in Montana State University's paleontology program. The man resembles a young Gary Sinise—bright smile, kind eyes, and fully attentive.

Originally from Southern California, Zach's lifelong interest in dinosaurs ensured he'd find his way to Montana—America's unofficial dinosaur capital. Colorado, Utah, and Wyoming might argue otherwise, but Montana makes a strong case for itself. When you take into consideration that Montana is home to North America's first recognized dinosaur remains (1854), the world's first identified *T. rex* (1902), North America's first baby dinosaur fossils (1978), and the best-preserved dinosaur ever discovered (2001), it's hard to dispute the state's dinosaur dominance.

"What drives you to study dinosaurs?" I ask Zach as he leads us toward the museum's Siebel Dinosaur Complex.

"For me, it's their diversity," Zach says. "They do so many strange things. From the ornamentation on their skulls to their habits, everything teaches us something."

Zach, like so many early career paleontologists in their twenties and thirties, is a member of the "*Jurassic Park* generation"—the phrase used to connect the 1993 blockbuster with the wave of paleontologists it inspired.

"I was actually born about a month after *Jurassic Park* was released," Zach explains. "And since my mom really wanted to see it, she watched it when she was still pregnant with me. I guess I was kicking a ton throughout the whole movie," Zach smiles. "She had to leave."

Zach leads us around the corner to the Hall of Horns and Teeth, where Ellie and I encounter our first true *T. rex* of the morning.

"Meet Peck's Rex," Zach says, nodding to the twelve-foot-tall, thirty-eight-foot-long specimen before us. "It's one of the largest and most complete *T. rex* ever found."

"Whoa," Ellie says, the only word she can manage.

"It was also one of the first *T. rex* specimens to preserve the vestigial third finger," Zach explains. For nearly eighty years, paleontologists had been uncertain whether *T. rex* had two claws or three. Thanks to Big Mike (whose bronze statue we admired out front) and Peck's Rex (now freeze-framed mid-roar before us), paleontologists were able to confirm that *T. rex* had two claws, along with a fleshy vestigial third, which over time receded both biologically and behaviorally.

The third claw debate might not seem important to humans of the twenty-first century. However, as the American Museum of Natural History notes, it proves that "even the best-studied of dinosaurs can still surprise us."

As surprises go, the *T. rex* still has plenty up its admittedly short sleeves. And as is true of all dinosaurs, we have much to learn through comparative biology. Zach explains that dinosaur fossils provide unique insight into life on Earth sixty-six million years ago. When we compare that insight with our current moment, patterns emerge. And within those patterns, we might learn new ways to improve our relationship with our planet.

"The biggest thing [dinosaurs] teach us is how organisms respond to change in their environment over time," Zach says. "We see how organisms respond to environmental stress or extinction. We see the range of adaptations organisms can have, which provides us with new information about animals currently alive and how we can help protect them and ecosystems today."

What Zach doesn't say—but what is abundantly clear—is that we humans are among those animals.

"As I understand it, human extinction is inevitable," I say, working through my logic. "And dinosaurs were around for over 165 million years, whereas we've barely even made it three hundred thousand. So, given how things are going," I reason, "it seems like they'll last a whole lot longer than we will."

"Well," Zach begins, "it depends on how you look at things. *Genera* of dinosaurs usually last a few million years, whereas *species* tend to last only a couple hundred thousand years before they either evolve into something else or die out."

"So . . . humans might just evolve into something else?"

"We might," Zach agrees. "Though I would argue that humans have kind of removed ourselves from the evolutionary chain. Any changes for us are by our choice."

Unlike other species, humans are not entirely at the mercy of the elements.

"We don't care if it's cold," Zach explains, "because we can always put on a coat."

I'm more concerned with temperatures fluctuating in the other direction: how it feels to shed everything but our skin and still be burning.

*

While strolling past a sign that asks, "What is paleontology?" I'm struck not by the answer ("the study of extinct life") but by the accompanying image: a black-and-white photo of a young Jack Horner.

Born in Shelby, Montana, in 1946, Jack Horner found his first dinosaur fossil at eight years old. It was a promising start for a man whose paleontological achievements would loom as large as the creatures he studied. Horner's best-known contribution occurred in the 1970s when he and a research partner discovered a new genus of dinosaurs called *Maiasaura*, which translates to "Good Mother Lizard." And for good reason.

After discovering several nests, eggs, and hatchlings, Horner and his dig partner, Robert Makela, hypothesized that *Maiasaura* cared for their young. The theory changed our understanding of dinosaur behavior, further dispelling the antiquated idea that dinosaurs were physically slow and dumb. Often credited as the inspiration for *Jurassic Park*'s protagonist, Dr. Alan Grant, Horner was also recruited as a technical adviser for several films in the franchise. Additionally, from 1982 to 2016, he served as curator of paleontology at the Museum of the Rockies, parlaying his star power and expertise to transforming a once-fledgling museum with eight dinosaur specimens into the thirty-five-thousand-specimen museum it is today.

Even in my short time loitering on the fringes of Montana's paleontological community, it's clear that everyone knows Jack Horner. In the weeks preceding our trip, I'd contacted him by various means, hopeful for an interview. I was greeted with silence, though I hadn't lost hope.

Even if, so far, it's been easier to unearth a dinosaur fossil than to track the man who's been doing it for decades.

*

Nearly 150 years after Edward Drinker Cope and Othniel Charles Marsh first faced off in the Bone Wars, a new twenty-first-century skirmish has emerged. Only now the battle lines are drawn differently. Whereas Cope and Marsh might best be characterized as men lusting for legacy, today's "adversaries" might be more easily split between academics and commercial fossil hunters.

The academics can be identified by their terminal degrees, university appointments, and access to federal grants. Commercial fossil hunters often lack all three. They're often passionate people with tools, time, and dig agreements with private landowners.

Academics adhere to the Society of Vertebrate Paleontology's code of ethics, which prohibits the sale of "scientifically significant vertebrate fossils" (except when they are sold to be brought into the public trust, such as to a museum). Commercial fossil hunters have fewer qualms about profiting off their finds. This is not to suggest that all fossil hunters are disinterested in the scientific value of their discoveries, just that they prioritize profits.

Thus, many academics view commercial fossil hunters as trying to cash in at the expense of science by selling to the highest bidder. Conversely, many commercial fossil hunters view academics as dismissive, exclusionary, and—due to their university connections—financially insulated.

However, the battle lines become further blurred when amateur paleontologists are considered. Since amateur paleontologists aren't professional paleontologists (read: they lack the degrees), they aren't academics. Yet many, like Tom Hebert, don't sell fossils, which means they aren't commercial fossil hunters, either. Where do we put the amateur paleontologists? It depends on who you ask.

Rancher and commercial fossil hunter Clayton Phipps—a.k.a. "The Dinosaur Cowboy" for his gallon-sized cattleman hat—has become a central figure in the ongoing debate between academics and commercial fossil hunters. In 2006 Phipps and a friend accompanied Phipps's cousin on his first-ever dinosaur fossil hunt. With the landowner's permission, they roamed a private ranch in the Hell Creek Formation, eventually stumbling upon what appeared to be a partially exposed dinosaur pelvis peeking from a hillside. Though interesting, Phipps's initial assessment was that excavating the pelvis did not make financial sense.

Nearly a month passed, and Phipps gave little thought to the find. But his cousin refused to let the matter rest. Phipps returned to the site at his cousin's urging, digging beyond the pelvis to discover a row of vertebrae and a ribcage. Two weeks into the excavation, Phipps had dug a perimeter around what appeared to be a nearly complete skeleton of a plant-eating dinosaur.

"I was working with a backhoe about five feet from the plant-eater thinking I'm completely safe because I had him completely laid out, I knew where the skeleton was," Phipps explained. "But just before I was about to dump my bucket, I noticed bone fragments."

Halting the machine, Phipps brushed away the dirt in the bucket.

"And then," he said, "this meat-eater claw starts to appear."

Then a leg, then an arm.

Phipps paused, wiped his brow, and took a moment for reality to sink in.

He hadn't found one well-preserved dinosaur but two. More astonishingly, judging by the placement of the fossils, the dinosaurs appeared to be locked in combat—their final battle neatly preserved sixty-six million years later. Better still, the combatants represented a royal rumble for the ages—a *Tyrannosaur* versus a *Triceratops*.

Removing the dinosaurs from the ground was relatively easy, but the accompanying public relations and protracted legal battles proved far more complex. Following the excavation, Phipps and his team shopped the Dueling Dinosaurs around to major museums but found no buyers.

In what felt like the next round in the academics versus commercial fossil hunters' saga, academics dismissed the importance of Phipps's discovery. "In order for a specimen to be of scientific use and publishable, we have to know its exact geographic position, its exact stratigraphic position, and the specimen must also be in the public trust, accessible for study, which this specimen is not," Jack Horner remarked.

In 2013, after years with no buyers, Phipps put the Dueling Dinosaurs up for auction at Bonhams in New York. While Bonhams's director of natural history played up the "dueling" aspects of the Dueling Dinosaurs—citing various tooth marks that seemed to support the interpretation—some paleontologists, including Horner, remained skeptical.

"They were found in river-channel sandstone, suggesting that the bodies were deposited on a sand bar," Horner said, adding that the odds of a pair of dueling dinosaurs purposefully taking their battle to a sandbar was unlikely.

"It's about sales," Horner concluded, "not science."

Many science-minded academics were troubled by the Dueling Dinosaurs' impending auction—which they feared would sequester the specimens into private hands.

Yet, in a twist, the bidding fell short of the $6 million reserve price.

Adding to the trouble was a dispute between the owners of the land on which the fossils were discovered. While one family owned the surface rights to the ranch and one-third of the mineral rights beneath the ground, the other two-thirds belonged to the land's co-owners.

Historically, fossils have belonged to those with surface rights, a precedent dating back to a 1915 Department of the Interior ruling stating that dinosaur fossils "are not mineral." Yet a 2018 U.S. Court of Appeals ruling for the Ninth Circuit upended that understanding, arguing that dinosaur fossils *could* be considered minerals. The decision imperiled not only the fate of Phipps's Dueling Dinosaurs (ceding ownership from one ranch family to another) but potentially all future fossil finds. While academics and fossil hunters didn't agree on much, they shared a mutual distaste for the ruling.

"Everybody in paleontology freaked out at that point," Phipps explained, "because no one had ever dealt with mineral holders. So any dinosaur that was ever from land that had severed mineral versus surface ownership was now under jeopardy of new ownership."

To minimize damage, in 2019 the Montana state legislature unanimously passed a bill confirming that dinosaur fossils are not minerals—much to the relief of paleontologists. However, the legislation exempted active lawsuits, leaving the fate of Phipps's Dueling Dinosaurs in limbo.

Two years later, the Montana Supreme Court reversed the Ninth Circuit's decision, freeing Phipps to pursue a sale.

"In my mind, a fifth grader could answer that an animal is not a mineral," Phipps said. "But it took seven years and a lot of money to settle that debate."

Rarely does a dinosaur become an albatross, but for Phipps, it did.

"I've lived my life trying to avoid attorneys and just be a decent human being," Phipps explained, "so it was an unfortunate thing we were caught up in. But," he added, "it turned out to be a good thing in the end."

Following the ruling, Phipps completed the sale to the North Carolina Museum of Natural Sciences, thus ensuring the specimens remain in the public trust, much to the relief of the scientific community.

Phipps, too, was pleased with the sale, which paid off both economically and scientifically. He was compensated for his work (Dueling Dinosaurs reportedly sold for $6 million), and scientists continue to have the opportunity to study the specimen. Despite the occasional critique from academics (some of whom have pegged Phipps as more interested in a paycheck than paleontology), the Dinosaur Cowboy remains no less committed to their shared endeavor.

"I've always hoped that the Dueling Dinosaurs would bridge the gap between people like me who go out and find dinosaurs and the academics who study them," Phipps says. "I find them so they *can* study them."

For many, Phipps serves as the poster child for commercial fossil hunters. Yet it's worth noting that he also serves on the board of the Garfield

County Museum, a stop on the Montana Dinosaur Trail, which, like all MDT stops, subscribes to the Society of Vertebrate Paleontology's code of ethics. Phipps's board membership may seem like the prehistoric elephant in the room, but the museum welcomes him, recognizing his support for the community.

Despite this service, some academics have trouble trusting Phipps for a different reason: he's a rancher whose formal education ended with a high school diploma. He lacks a proper paleontology degree, but then again, so did Edward Drinker Cope, who received an "honorary" master's degree instead.

While the dirt doesn't care if you've earned a degree, professional paleontologists often do. It's not just academic snobbery; professional paleontologists dedicate years to honing best practices and procedures for fossil extraction and data collection. They want others to adhere to the same high standards for science's sake.

Phipps works in a manner that he hopes will silence the skeptics. He's continuously refining his techniques and learning by way of experience. And it's not exactly his first day on the dig site. Phipps hunted fossils for two decades before finding Dueling Dinosaurs. He did so to make a living from the land—whose rugged environment had proved less than ideal for farming and ranching.

In 2013 Phipps told a reporter that his 1,100 acres were "a little too big to starve to death on." Selling fossils proved a new way to feed his family when he desperately needed the cash.

"Some years," Phipps says, "I was doing this while living hand to mouth."

Less so these days. In addition to selling the Dueling Dinosaurs, he's also the star of the Discovery Channel reality show *Dino Hunters*. (I suspect the show's made him few friends in the professional paleontology community, given its description, which emphasizes "buck[ing] the academic status quo.")

It's the kind of mainstream exposure that few academics ever receive. What Phipps lacks in formal paleontology training, he makes up for with hard work and a TV-ready personality. And he knows what all fossil hunters must: finding fossils takes time, commitment, and persistence.

"There's no roadmap to finding a dinosaur," Phipps says, "but you're not going to find one sitting on the couch."

*

As we wind our way toward the end of the Hall of Horns and Teeth, I ask Zach the question most weighing on my mind: Is there any middle ground between academics and commercial fossil hunters?

Zach pauses to give it some thought.

"You know," he says, "when an important specimen is found, casts and scans can be made of that specimen. Then those can be provided to museums [before a private sale]. You wouldn't get all the information this way, but at least you could learn a little."

Though imperfect, it's at least close to a compromise—using technological advances within the field to provide greater access to all. Yet when it comes to sharing information, greater access to all isn't always the goal. In paleontology, limiting access also has its uses—especially when it involves dig site locations and proprietary information. Such secretiveness—though often well intentioned—complicates collaborative research, even between academics. Add an additional layer of distrust between academics and commercial fossil hunters, and "greater access to all" might mean "greater access to all with the proper credentials."

While the relationship between academics and commercial fossil hunters may seem adversarial, Zach believes there's room for compromise. The devil is in the details, or in this case, the dinosaur species involved.

"If you're working on a *T. rex* or a *Triceratops*, well, there are lots of *T. rex* and *Triceratops*," Zach explains. "In the grand scheme of things, if you miss one, that's probably not a big deal. But a *Pterosaur* find would be much more rare. And when it's rare," he concludes, "we need to have science involved."

*

Upon leaving Bozeman's Museum of the Rockies, we head northeast toward Harlowton, a small town in Wheatland County with a population under a thousand. But halfway to our destination, Ellie grows hungry.

"We've got snacks in the back," I try.

"But I'm starving," she says. "Can't we stop somewhere for a real meal?"

"Somewhere where?" I ask, eyeing the emptiness ahead.

Eventually, we spot a town called Big Timber, where we find the Grand Hotel and Restaurant—a sturdy black brick structure that looks built to last. In fact, its history proves it. Originally constructed in 1890, the Grand Hotel was a regular stop for sheep ranchers and travelers on the Northern Pacific Railroad. In 1908 a fire destroyed most of the town's businesses, but the Grand Hotel remained.

Today the restaurant and hotel are a favorite haunt for several ghosts—all of whom we are introduced to thanks to our waitress, Josey Montana. No sooner do we sidle up to our stools than Josey—our sixteen-year-old server—takes one look at us and determines we are "ghost people."

"This place is *super* haunted," she says with little prompting. "There are actually three ghosts here."

"Really?" Ellie asks, suddenly forgetting her near brush with starvation. "Can you tell us about them?"

According to Josey, the first ghost is a Chinese laundry woman who worked in the hotel a century ago.

The second, she continues, is a nondescript woman regularly spotted in room 6.

The third, Josey concludes, is a young girl about Ellie's age with blonde hair.

"I've only been here since October," Josey says, "but those stories are very, very true. Sometimes, trays flip out of our hands, and glasses randomly shatter. Sometimes, we close off certain tables for the night. Someone doesn't want us sitting there."

Our eyes widen.

"But room 6 is where the real action is," Josey continues. "People sometimes feel dizzy or chilled. And the toilet flushes by itself. While you're waiting for your food," she says, "feel free to sneak up the stairs and check it out yourself."

If it's a dare, we take it. Or at least I do, reaching for Ellie's hand moments after we order.

"Are you sure we're allowed up here?" Ellie asks as the stairs give way to a long, dark hallway.

"Would Josey Montana lead us astray?" I ask.

I tiptoe toward room 6 while Ellie hangs back at a safe distance. I am in full ghost-hunter mode—an ear cocked toward the ceiling as I await a phantom flush. Though we are safely on the opposite side of the door, it does little to ease Ellie's fears. What ghost worth its salt has ever been stopped by a door?

"Wait . . . do you hear that?" I ask.

"Hear what?" Ellie whispers.

"Silence," I whisper, grinning.

She is not impressed. Nor dizzy or chilled. Mostly, she's just back to being hungry.

Equal parts disappointed and relieved, we retreat down the stairs to find Josey delivering our salad and sandwich.

"Any ghosts?" she asks as we return to our stools.

"Not this time," I say. "Not even a phantom toilet flush."

"Well, maybe next time," Josey smiles. "Enjoy your food."

In one swift motion, she hands over the check.

Dinosaurs, after all, aren't the only way to make a buck in Montana.

*

We arrive in Harlowton by midafternoon, bypassing several boarded-up buildings along an empty street. There are few people anywhere. Yet, like many small Montanan towns, they are just one *T. rex* discovery away from kickstarting their economy.

Ellie and I enter the Upper Musselshell Museum—our second stop on the Montana Dinosaur Trail—to find Tim Busby, the bearded, bespectacled museum director, sitting behind the front desk.

"Well, hello there," Tim greets us. "Welcome to town."

I introduce us, explaining that we're the ones working on the dinosaur book.

"Ah, yes," Tim says. "Well, right this way. Allow me to introduce you to Ava."

If they gave prizes for the cutest dinosaur, Ava—formally known as *Avaceratops lammersi*—would be a clear winner. *Avaceratops* is a member of the *Ceratopsian* family, and it resembles a smaller version of a *Triceratops*. Measuring seven and a half feet long, this one is about the length of a Komodo dragon. At full size, it might've grown nearly as large as an African rhino. Beaked and frilled, it lived in a leafy-green forested environment, ideal for a herbivore. And Ava died there, too, seventy-six million years ago, just twelve miles east of where we now stand.

Today, the museum exhibits a cast of Ava's fossils. (The Academy of Natural Sciences in Philadelphia is currently studying the authentic fossils.) Yet even the cast is awe-inspiring.

Though *Avaceratops* and *Triceratops* share a family, they're more like cousins than direct descendants. "For one thing," Tim says, "*Triceratops* is from the Cretaceous period but lived more than three million years after Ava. When Ava lived there were around twenty-six different species of *Ceratopsians*, but by the time of *Triceratops*, there were only a few species of *Ceratopsians* left. A second difference is the horns. You see the veins here? She had keratinous horns, like a cow. So her horns would continually grow. But they were a sheath rather than a solid horn, whereas *Triceratops* had solid bone horns."

Ellie and I nod along bobble-head style.

"The last difference," Tim continues, "is Ava didn't have teeth. Because the grass had not evolved until late in the Cretaceous, she was eating woody plants, ferns, and things like that, so her jaws are more like garden shears."

Tim can—and does—go on. Though he has no formal training in paleontology, his extensive knowledge of dinosaurs, geology, and state history is humbling. As I'll soon learn, it's not uncommon for small-town museum directors to know a bit of everything and then some.

Though Harlowton is a long way from Times Square, or even Billings, Tim values the town for what it is: a peaceful place with plenty of beauty and not too many people. It's got an important story, too, which might begin when Ava first roamed this land and continue right into the present. It's a story of dinosaurs but also of Indigenous tribes, including the Crow, Blackfeet,

Flathead, Gros Ventre, Northern Cheyenne, Nez Perce, Shoshone, Sioux, and Assiniboine. And it's the story of a railroad that helped build the town before leaving in the 1970s—taking 20 percent of the population with it.

Tim says the Montana Dinosaur Trail provides an entry point into these stories, offering wider exposure to Harlowton and many other often-overlooked towns along the trail.

"The Montana Dinosaur Trail brings people to places they might otherwise drive through," Tim says. "It raises people's awareness that we're even here."

*

We drive east for an hour, rolling into the city of Roundup shortly before nightfall.

"Well, look who it is," Tom Hebert calls, "the world travelers."

We hadn't seen him since the previous afternoon, shortly before we parted ways near Miles City after finding the *Hadrosaur* bone. Now Tom's staffing the table for Earth Sciences Foundation at the city's biggest event of the year—the Roundup Independence Day Extravaganza (colloquially called RIDE), where most of the town's 1,800 or so residents have descended for three days of inflatables, carnival rides, and more huckleberry lemonade than anyone needs.

Adding to the festivities are Tom's wife Beth and their eleven- and thirteen-year-old nephews, who join Tom in the event tent.

"What do we got here?" I ask, nodding toward the table overflowing with chalky white casts.

"We're just opening some plaster jackets to see what dinosaur fossils are inside," Tom explains.

Plaster jackets (also known as field jackets) are white plaster bundles wrapped around fossils to ensure their safe transport. These jackets are from Fort Yates's Standing Rock Institute of History, and Tom assures us that all the fossils will be returned there.

Crowds wander toward the tent in twos and threes, their interest piqued by Tom's "jacket unveilings"—most of which contain an assortment of *Had-*

*rosaur* material. The fossils were collected years ago on Standing Rock Sioux land by students at Sitting Bull College. As their paleontology program dried up, so did the opportunities to continue their fossil prep work. They left behind hundreds of plaster jackets, a few of which Tom opens now before his live audience.

For Tom, it's just another day at the "office," but for festivalgoers, some of whom are seeing dinosaur fossils for the first time, it's the night's highlight. No matter that a Styx tribute band is blasting "Mr. Roboto" on stage just ahead of us; here, there are dinosaurs.

A few onlookers reach into their wallets and drop a dollar or two into Tom's donation "jar"—a five-gallon water drum that could use a few more dollars. Dinosaurs are big business, but for foundations like Tom's, they're small business too. Tom doesn't mind the fundraising hustle; he's made peace with the fact that the money will come or it won't. Since it hasn't yet, he hasn't cut himself a check for his foundation work. Grants and donations subsidize the foundation's other expenses, as does the occasional fundraising raffle. At this year's RIDE, Tom's raffling off a pair of football tickets for Montana versus Montana State. However, not many people feel lucky at fifty dollars a raffle ticket.

In a previous fundraiser, Tom raffled a *T. rex* tooth. The tooth was found on private land and donated to the foundation. By raffling it, Tom hoped to raise a few hundred dollars to support a free dinosaur dig for kids. In doing so, he instead raised the ire of at least a few in the paleontology community, who felt that raffling a *T. rex* tooth—while not technically "selling" a fossil—was dangerously close to violating the Society of Vertebrate Paleontology's code of ethics, which states that "the barter, sale or purchase of scientifically significant vertebrate fossils is not condoned."

While Tom remains unsure if a single *T. rex* tooth qualifies as a "scientifically significant vertebrate fossil" (for the right price, you can pick one up at the nearest rock shop), Tom took the criticism seriously and never raffled a fossil again.

Tom's regularly assured me that he's never sold a dinosaur fossil and strictly adheres to the society's code of ethics. But when he raffled the tooth, the damage was done. If the academic community was looking for a reason to dismiss Tom Hebert, he gave it to them.

"If I could do it over again," Tom says, "I wouldn't have done it."

While the code of ethics is well intentioned, what happens when a museum purchases a "scientifically significant vertebrate fossil" to keep it in the public trust?

This was the case in the auctions for Tyrannosaurus Sue and the Dueling Dinosaurs. Neither museum sale was ideal (both involved buying scientifically significant vertebrate fossils, after all), but since they were sold to keep the specimens within "a public trust," both sales complied with the code. Were the museum's purchases not the lesser of two evils? Isn't it better to sell a dinosaur to a museum to be enjoyed by all rather than to a millionaire to be used as a coatrack?

Amid all the nuance, one thing is clear: trouble's never far behind when it comes to buying and selling dinosaur fossils. Selling fossils is illegal in some instances and ethically fraught in others. Yet for commercial fossil hunters with the proper permissions, it's simply a way to earn a living.

In some ways, Tom's a man adrift in the world of paleontology. He's neither a commercial fossil hunter nor a professional paleontologist. As an amateur paleontologist, he leans toward the latter, though few professional paleontologists have warmed to him. While he's regularly invited to present at geology and paleontology conferences nationwide, by his own admission, he has yet to receive the scientific credibility he seeks.

At the Roundup Independence Day Extravaganza, Tom's main objective isn't to win over scientists but to make a few more friends for himself and his foundation. After all, Roundup has become his newly adopted home. It's a long way from Wisconsin, but Tom doesn't mind the drive. Particularly because he has big plans for the city. One day, he hopes Roundup, Montana, will become the world headquarters for Earth Sciences Foundation. In addition to that, Tom also wants to assist the city's Musselshell Valley Historical Museum in securing a coveted spot on the Montana Dinosaur Trail. If Roundup can be the home to the foundation's headquarters and

# 4
# Tires Make Good Neighbors

## JULY 2
*Roundup,* MT

"Well, this ain't good," Tom sighs, eyeing the back flat tire of his pickup truck. "Second one this week."

As flat tires go, Tom picked a good place to get one—on a ranch outside Roundup, adjacent to the guest house where we'd stayed the previous night. And where we'll stay the next night, too, thanks to the generosity of the ranchers who'd loaned Tom the house to support his dinosaur digs. We'd been prepared to pitch a tent, though western hospitality wouldn't hear of it. In the guest house, there's room for us all.

The previous night, while driving from the Roundup Independence Day Extravaganza to the ranch, Ellie and I caravanned behind Tom as dusk gave way to darkness. We drove until the paved roads turned to gravel, then dust. I stayed just a vehicle length behind him. There wasn't much chance of losing him, but if I did, there wasn't much chance of finding him again, either, amid the unspooling terrain.

Back home, I'd grown accustomed to the Midwest's grid-like sequences of farms and fields—the interlocking pieces of a vast puzzle. But the West refuses any such containment. It is a study in contrasts: grasslands to mountains, prairies to peaks. Out here, the land determines its own course.

Outside our bug-splattered windshield, the expanse seemed endless. As darkness descended, the clouds broke to reveal an ocean of stars.

Driving, I was reminded of a line from comedian Rich Hall: "In Montana, a policeman will pull you over because he is lonely."

A sentiment I suddenly understood.

When we arrived on the ranch's fifteen thousand acres, it was too dark to see any of it. Only now, in the brightness of morning, as I peer down at Tom desperately trying to fit the tire iron to the stripped lug nut, do I begin to understand just what fifteen thousand acres means.

It means twenty-three square miles—nearly two-thirds of our city back home. All this land is dedicated to five hundred or so very fortunate Angus cows and calves who graze upon the seemingly never-ending acreage.

"Normally, you don't have to pound on 'em like this," Tom says, taking a wrench to the lug nut. "But this one," he grits, "just doesn't want to give."

As a card-carrying AAA member, my knowledge of tire-changing begins and ends with the phone number on the back of the card. I imagine such pronouncements will earn me no street cred here in Montana and even less with Tom, who never saw an obstacle he couldn't overcome with dogged determination—and, in this case, brute force applied to metal on metal.

"There it goes," Tom grunts, freeing the last lug nut. "Now we're in business."

There's still the problem of finding the perfectly sized replacement truck tire at the last minute in a town of 1,800, but Tom assures me he has a few good leads.

"I put it out on social media," he says—a strategy that yielded four offers for loaner vehicles and the tire itself within an hour or so.

"Do you know these people?" I ask. "The ones offering help?"

"Some. But not the guy with the tire," Tom says. "He's on vacation. He said to go into his trailer and help myself."

I lift an eyebrow. "Some stranger you've never met told you to go into his trailer and steal a tire?"

"Not *steal*," Tom explains, depositing the lug wrench into the truck bed. "*Borrow*. Once I drive to Billings and pick up a new tire, I'll put the old one back where I found it. Chances are I'll never meet the guy who's loaning it. He's on vacation till next Friday."

The "enter my home when I'm not there" approach juxtaposes starkly with the deadbolts and doorbell cams back home. I feel safe in Wisconsin, but I still lock the doors. And I've never invited strangers to rummage through my stuff when I'm away.

"You've got some pretty friendly neighbors here in Montana," I say.

"Yeah, well, we've learned to depend on each other," Tom says, peering out at the morning. "Out here, it's the only way to survive."

*

Back in the guest house, Ellie, Tom, and Beth's eleven- and thirteen-year-old nephews lounge on the couches while Beth fries a package of bacon.

"Morning, sweetheart," Tom says.

"Morning," Beth says, turning away from the pop and sizzle of the bacon grease to meet her husband for a kiss. "I'm looking for paper towels. Think you can help?"

"I've got a solution for every problem you have," Tom says, scouring the cabinets.

If there's one thing Tom loves more than dinosaurs, it's Beth. Together, they provided one another with structure, support, and companionship when they needed all three. Following his 2012 divorce, Tom entered a self-described "downward spiral."

"I was a real ass for a long time," he explained.

Through it all, Tom remained sober, though he struggled with frequent bouts of anger, depression, and the occasional irresponsible shopping spree. He spent his days chained to his insurance agency, work that provided a paycheck but little else. It wasn't his lowest point—he'd hit that while seated with his handgun at his kitchen table in September of 2002—but it was close. And there were no signs of improvement.

Until he met Beth.

Following a yearlong courtship, they got engaged in 2019, then married in 2020—just months before Tom enrolled in my class.

While Beth is Tom's greatest supporter, she's not the only one. Earth Sciences Foundation exists due to its unique coalition of supporters: the Standing Rock Sioux Tribe, for starters, and several Montana ranch owners who've permitted Tom to dig on their land. But also, a commercial property owner in Roundup who sold the ESF its future headquarters at a bargain price—prime real estate on the downtown strip. Help has also come from a regional philanthropic foundation that awarded the ESF a major grant. And from members of the geospatial tech industry who've loaned Tom tens of thousands of dollars' worth of surveying equipment.

But for every supporter, there's a skeptic who questions Tom's intentions, integrity, and paleontological knowledge. The rumor mill swirls with allegations of improprieties, though Tom has been candid with me about his missteps.

In addition to raffling the *T. rex* tooth, Tom ruffled more feathers within the paleontology community when he referred to himself as a "paleontologist"—a designation a few in the field took umbrage with, believing he was attempting to pass himself off as a "professional paleontologist." According to the American Geosciences Institute, professional paleontologists must have obtained an advanced degree (a master's or a doctorate), whereas Tom is still working toward his bachelor's. Though he no longer uses the title, Tom says he didn't mean to misrepresent his credentials ("I dig up dinosaurs; that's what paleontologists do"). Had he clarified that he is an "amateur paleontologist," there might not have been an issue. However, his failure to specify the precise designation further hindered his acceptance into the professional paleontology community. (It's worth noting that in all my conversations with "professionals" and "amateurs," I've never heard anyone make the designation.) Today, when featured in a podcast or an article, Tom employs an indisputable title: director of Earth Sciences Foundation Inc.

Tom's lack of specificity regarding the paleontologist title wasn't his first unpleasant encounter with academics. Even as an undergraduate student, Tom often felt entangled in the bureaucracies of college life. In the spring of 2021, Tom and three fellow students presented their land surveying research at the Geological Society of America's national conference, though back at the university, he struggled to garner the attention of his professors.

Feeling invisible in his own backyard, Tom transferred to another university shortly after our class. He might have dropped out if not for a late-night call in the spring of 2021, during which he and I discussed some of his academic struggles.

"Tom, you can't quit," I said, roaming my backyard, my cell phone clutched to my ear. "From everything I'm hearing, you're doing great."

But Tom had grown restless. His problems were less about any particular class or professor than higher education itself, which, to him, felt like jumping through one hoop after another with the simple trust of a well-trained seal.

He was forty-four and not getting any younger. He wanted to earn his geology degree and discover dinosaurs, but his plan had proved challenging. Perhaps what he wanted most was a mentor in the scientific community, someone who might've encouraged him not to raffle the *T. rex* tooth or refer to himself as a paleontologist without specifying which kind.

"Tom has this dream of making it possible for every person to truly access the world of science," Beth explains as Tom reaches into the fridge to retrieve a carton of eggs. "And he wants to do it in a way that won't cost people a million dollars. Or exclude anyone because they don't have the right degree. Or because they don't know the right person."

"You know that song in *Hamilton* where Aaron Burr sings about wanting to be 'in the room where it happens'?" Tom asks.

I nod.

"In the paleontology world, I will never be in that room because I don't have the alphabet soup of degrees behind my name. And because the people who do have the alphabet soup behind their names are the gatekeepers," Tom says.

As someone with at least a little "alphabet soup" after his name, I can't help but feel a bit complicit. As a professor who's taught well over a thousand students, I like to think I know a thing or two about the challenges undergraduates face. I know college is a significant investment that often saddles students with debt for years. I know, too, that some classes can feel nebulous, and some professors can be mercurial—the combination of which can leave some students feeling at the mercy of a system they don't fully understand. But I also know that college opens more doors than it closes. And that for many students, "the room where it happens" is where a professor and students take their seats around the table to learn and grow together—much as Tom and I had in our rhetoric and composition class. However, as Tom is quick to point out, college is just one form of gatekeeping. The broader scientific community can occasionally feel a little exclusive too.

"I'm not saying you let everyone into the room," Tom continues. "But we have to be open-minded enough to have more than one way in. Doors, windows, take a chainsaw and cut a hole through the wall if you have to. We have to respect all those ways of getting in."

*

Established in 1883, Roundup was—as the name suggests—the epicenter of the West's cattle roundups. Tucked along the shores of the Musselshell River, the open range was a perfect place for stockmen to sort and brand cattle before shipping them east.

The city's visitor bureau reports that these roundups were most successful when stockmen worked together, "relying on the honor system" to ensure the cattle were accounted for and that payments reached the right hands.

"Stockmen had plenty of other reasons to cooperate—wolves, coyotes, rustlers and newcomers all threatened their cattle," the visitor's bureau explains. "No one could survive without the help of fellow ranchers."

That spirit seems to have endured, at least as tires are concerned.

After Tom borrows a truck to pick up and install the tire, we caravan back into town. We pass the gas station and the grocery store before parking on Main Street, in front of a two-story shuttered building amid a row of shops.

"Welcome," Tom says, inserting a key into the lock and giving the doors a good hip check, "to the future home of the Earth Sciences Foundation!"

Beth and The Nephews enter while Ellie and I bring up the rear.

Built in 1911, the building has fulfilled many roles within the community for the past century: from bank to post office to general store. Within a year, Tom hopes (after greasing the hinges) to fling the doors wide yet again, this time to showcase the region's dinosaurs at the foundation's first headquarters.

"Most recently, this place was called Evans Gifts," Tom says. "They closed in 1989. And as you can see, it hasn't really been touched since."

Part time capsule, part hoarder's paradise, the building is chock-full of history. Sunlight streams through the dust-caked windows, revealing stacks of boxes, piles of bags, and a spinner rack overflowing with greeting cards for every occasion. My eyes settle on a box of 1980s-era NBA basketball buttons. To our left, I spot several board games older than me. Near the back, a shelf overflows with yellowed newspapers, magazines, and books ranging from Mary Shelley's *Frankenstein* to Woody Guthrie's *Bound for Glory*. Opposite the shelf, and keeping with the eclectic theme, is a framed portrait of Nez Perce leader Chief Joseph.

*This place is already a museum,* I think.

Ellie and The Nephews explore the gift-shop-turned-cabinet-of-curiosities. Each display case and bin has the potential to hold some treasure—*T. rex* femurs, Spanish galleons, your guess is as good as mine. Ellie's eyes widen as she and The Nephews brush dust from various boxes, flipping wide their cardboard flaps and peering at the wonders within. I lose sight of her within seconds as she presses deeper into the head-high stacks. For a moment, there is no sign of her except for the squeak of the century-old floorboards beneath her feet.

"From this counter to here, this will all be a gift shop," Tom says, giving me a tour of what he hopes this place will become. "Now, from here back," he continues, "this will be our museum display area—rocks, dinosaur fossils, that sort of thing. And in the far back, that will be our conference room, where we'll do virtual presentations."

The foundation hopes to convert the upstairs into a natural history exhibit featuring animal mounts and art. Meanwhile, the basement will serve as a fossil prep lab, a hands-on experience beyond the field.

"So what's the next step?" I ask.

"The next step," Tom says, eyeing the mess, "is to clean everything out."

"But there aren't enough hours, money, or hands," Beth says, peering into a cardboard box.

"Is Roundup willing to help you bring this to life?" I ask.

"It kind of depends on who you ask," Tom says. "Just yesterday, the mayor came up to me and gave me a big hug and said, 'When are you coming to city council to talk about what you have to do?' But then I'll try to work with the museum—where I helped install an entire dinosaur exhibit free of charge—and instead of thanking me, I'll get phone calls about how I need to install plexiglass because some dirt from the exhibit is getting on their floor."

Tom's referring to the Musselshell Valley Historical Museum, an independently owned and operated museum that, since its founding in 1973, has helped tell Roundup's story—from cowboys to cattle drives to coal mines. Thanks in part to Earth Sciences Foundation, the museum now has a room dedicated to dinosaurs—complete with various fossils on loan and an artistically rendered *Triceratops* mount.

Initially, the foundation's partnership with the museum seemed promising. However, over the past several months, their relationship has become strained. The museum and foundation originally planned to work together to earn Roundup a coveted spot on the Montana Dinosaur Trail. However, there are growing concerns that Tom's affiliation with the effort might derail the museum's chances of success.

Tom blames politics and personalities, for which he is quick to concede that he is partially responsible.

"I've been told I don't always play well with the other kids in the sandbox," Tom says. "When I get frustrated, I have a tendency to make bad decisions that make a bad situation worse."

Truth be told, Tom's been known to butt more heads than a dome-skulled *Pachycephalosaurus*. He knows, too, that rectifying his relationship struggles likely involves adopting a humbler approach.

Tom isn't winning any popularity contests within Montana's paleontology community, and he has no delusions about how he's perceived.

"There are people out there," he says colorfully, "who wouldn't piss on me if I was on fire."

Ellie and The Nephews reemerge at the far end of the building, having unearthed several brown travel mugs with "Evans Drugs" stenciled along the side. They flip through fifty-year-old books and peer into drawers of a filing cabinet so tall that it nearly scrapes the ceiling. It's a fun place to explore, but transforming it into the foundation's headquarters seems like a heavy—and costly—lift.

"Are you sure it's worth it?" I ask, peering out at the cluttered room. "Especially given the city's lukewarm response?"

"It's worth it," Beth chimes in, "because of the people supporting us. Our struggle is that we fell in love with this town and the people here. And we've wanted to do everything we can to help and be a part of this community. We've even talked about moving here full-time. But then you see the character of certain individuals," Beth says, "and it makes you question what you think you know. As you've witnessed, this is a place where a stranger will help you with a flat tire. But then you've got other people . . ."

As evolution has taught us, change takes time. And when personalities are part of the problem, people can be slow to change.

*

Late in the afternoon, after completing one too many Mad Libs, Ellie and I fight about adverbs while seated beneath a tree in the park adjacent to the midway for the Roundup Independence Day Extravaganza.

"Look, adverbs are easy," I say. "Just think of words that end in *-ly*. Dreamily, studiously, angrily . . ."

"Angrily . . ." Ellie growls. "I choose 'angrily' . . ."

"Honey, it's just a game. You did great with the nouns . . ."

"I'm not stupid," she says.

"Nobody said that you were."

"Well, you think it!"

"I really don't!"

"We can't all be English professors, Dad!"

I sigh.

Adverbs were not high on the list of what I expected us to fight about this trip. Though, of course, we're not really fighting about adverbs. It's hot, we're tired, and we're a long way from home. At least one of us misses our mommy. Also, five days of near-constant motion have started to wear on us. Though we aren't acting like it, we're grateful for our day of rest in Roundup. It's a beautiful afternoon in a beautiful town, and there's no shortage of elephant ears or huckleberry lemonade.

"You know what? Let's forget the Mad Libs," I say, tossing the book and reaching for the deck of cards. "How about a new game?"

"What?" she asks, her sourpuss face slowly giving way to curiosity.

"You're going to love it," I say, dealing the cards.

A hundred yards away, Tom, Beth, and The Nephews work the Earth Sciences Foundation table, entertaining onlookers to today's game of "What's in the plaster jackets?" Ellie and I had assisted for much of the blistering afternoon, though eventually, we retreated to the shade tree in the park.

I finish dealing.

"Okay," I say, "look at your cards."

Ellie does.

"Got any threes?"

"No."

"Fours?"

"No."

"Then draw four and discard one."

This goes on for some time—I make up rules, and she follows them.

"Do you have any runs?"

"What's a run?"

"If not, Go Fish!"

"Go . . . Fish?"

"Old Maid wins!"

"Huh?"

By now, the game is near madness, but suddenly, Ellie is all in. Anything beats adverbs—especially a four of spades.

She laughs as she splays several cards in the grass for no reason. Suddenly, we are having the time of our lives.

"A six-five combo!" I shout. "Two points for you! Now we gotta shuffle . . ."

As Ellie attempts to devise some strategy for our made-up game, I take a moment to survey my surroundings—inflatables, carnival rides, food vendors, a pool, a playground, and a thousand or so milling-about Montanans, their smiles as big as their belt buckles.

It is a happy scene, and we are happy to be a part of it. And privileged, too, to have been momentarily welcomed into this world of the West. Somewhere in this crowd are the folks who offered Tom a loaner vehicle. But so are the people who wish Tom would pack up his dinosaurs and leave.

I return my attention to my cards.

"Uh-oh."

"What?"

"Read 'em and weep," I say, revealing my random sequence of cards—a 2, 4, 7, and a Jack.

"What's that mean?" Ellie asks.

"Eleven points for Dad."

"What?" she asks. "But . . . how?"

"Honey, that's just the game," I say. "I didn't make the rules."

"But you did!" she laughs. "You made all of them! I mean, how can I win if I don't know the rules?"

My smile weakens.

"Well, hon," I say, peering as Tom cuts into a plaster jacket in the distance. "I'm afraid that's life."

*

Four huckleberry lemonades deep, Tom is more huckleberry than man.

"Maybe you want to coat your stomach with a little food?" I suggest, standing alongside the tent.

"Hey," he says, lost in the throes of his huckleberry high, "huckleberry *is* a food."

It's a few minutes before 8:00 p.m. A crowd of at least a thousand has gathered at the outdoor stage, anxiously awaiting Sawyer Brown. The tech guy is doing a mic check ("Check, check, check, 1, 2 . . .") when Tom—between slurps—glances at his watch.

"Oh shoot," he says, placing his cup on the table, "I'm supposed to be on stage!"

"Wait . . . what?" I ask.

Tom takes off toward the stage, and Ellie and I follow in hot pursuit, weaving through the crowd until we reach the front.

Though I wouldn't be entirely surprised to learn that Tom sings backup for Sawyer Brown, it seems a bit of a stretch. This makes me even more curious about why Tom is preparing to take the stage.

"Good evening, everybody!" the emcee calls to the crowd. "How are we all doing tonight?"

A whoop goes up from the crowd in their foldout chairs.

"Now, before we get to the music, we've got a couple of very special awards to share with you." The emcee calls a representative from the region's energy company's foundation to the stage and introduces the year's grantee award recipients.

Including Tom Hebert, director of Earth Sciences Foundation Inc., who makes his way to the stage when his name is called. The crowd—more interested in Sawyer Brown than in Tom—politely claps. Tom smiles, posing with his plaque as the photographer snaps a photo.

Back at the ESF tent, a beaming Tom reunites with the rest of us.

"Congratulations on your award," Beth says, kissing him.

"Forget the award," Tom laughs. "I just met Sawyer Brown! Backstage. I gave my card to their manager. I told him I'd take the whole band out on a dig. Maybe we can work it in before they leave town!"

I smile.

It is the most Tom Hebert thing Tom Hebert's ever said.

Only in his world could it maybe happen.

*

FIG. 9. B.J. and Ellie await the Sawyer Brown concert at the Roundup Independence Day Extravaganza in Roundup, Montana. Author photo.

Sawyer Brown strums and sings long past sundown, playing all the hits. I've never heard most of them, though the crowd has, the lyrics seemingly sequenced into their DNA. Peering out at the sea of people stretched along the park, I observe expressions ranging from pride to wonder. Both of which seem spot on. After all, bringing a beloved national act to a small town in central Montana is no small thing. And they've done it. And they're proud. If a city's good enough for Sawyer Brown, then a city is good enough.

The night turns patriotic on the eve of the eve of the Fourth of July. As the Sawyer Brown set comes to a close, a recording of Lee Greenwood's "God Bless the USA" pipes through the speakers as fireworks light up the night.

Ellie, The Nephews, and I lean against the fence just a stone's throw from the stage. Our necks tilt high as the fireworks boom like cannon fire. At the Earth Sciences Foundation tent, Tom beams brighter than the fireworks. Several Roundup friends gather around him and Beth. For the moment, all hostilities between Tom and the town seem to have dissipated.

A crescendo of cheers rises and falls throughout the park as the last fireworks dim. Every person in the crowd leaves with a smile, a dreamy-eyed parade of Sawyer Brown–loving patriots returning to pickups and cars.

Many of these folks have a drive ahead of them. But none have come farther than us.

As Ellie and I walk back toward the tent, we receive more waves, hellos, and tipped hats than in the rest of my life combined. Warmth radiates from the crowd as mothers and fathers hold their children's hands. Laughter is the soundtrack for every step along the way.

"Well, Tom," I say, "if it's all the same to you, I think we'll meet you back at the ranch. We got one who's getting pretty tired over here." Ellie yawns to confirm it.

Tom pauses from packing up the plaster jackets and turns toward us.

"You sure you know where you're going?" he asks.

"Not exactly," I admit.

"Well," Tom chuckles, slapping my back, "if you run into trouble, chances are someone will stop."

FIG. 10. Charles H. Sternberg searches for fossils in Ellsworth County, Kansas, in 1906, nine years after the death of Edward Drinker Cope. Sternberg, George Fryer 1883–1969, "113-02: Charles H. Sternberg" (1906). *George Sternberg Album #1 – Early Pictures*. 507, Forsyth Library Special Collections, For Hays State University.

# 5
# Where the Deer and the Dinosaurs Play

## JULY 3
*Roundup, MT*

There was never a time in Charles Sternberg's long life when he was ever far from a fossil. Born near Cooperstown, New York, in 1850, Sternberg spent his early years exploring the region's wilderness in search of fossilized plants and shells. At seventeen, Sternberg committed himself to making his hobby his profession. "I made up my mind what part I should play in life," Sternberg wrote, "and determined that whatever it might cost me in privation, danger, and solitude, I would make it my business to collect facts from the crust of the earth."

Despite his reverend father's inability to see the "practical side" of digging in the dirt ("He told me that if I had been a rich man's son, it would doubtless be an enjoyable way of passing my time"), Sternberg left his then-home of Iowa to join his older brother in central Kansas—a hotbed for fossilized plants.

Sternberg's hands soon became divining rods. For days, he'd roam the hills of Ellsworth County, returning home with collecting bags overflowing with fossils.

By 1870 he'd collected such an impressive array of plant fossils that he sent some of his findings to the Smithsonian Institution for further study.

"There was no money in fossils at that early day," Sternberg wrote, "but I prized more highly than money the promise in the [Smithsonian's] letter that my specimens would be studied by competent authority, and that I should receive credit for my discoveries."

Though Sternberg seemed to have momentarily earned himself a seat at the scientific table, he could hardly exist on credit alone.

In 1875, during his year at Kansas State Agricultural College (known as Kansas State University today), Sternberg attempted to secure a spot on a fossil-collecting expedition led by Professor Othniel Charles Marsh of Yale. Despite his efforts, all the spots were filled.

Sternberg reached out to Marsh's rival, Professor Edward Drinker Cope.

"I put my soul into the letter . . . for this was my last chance," Sternberg later remarked. "I told [Cope] of my love for science, and of my earnest longing to enter the chalk of western Kansas and make a collection of its wonderful fossils, no matter what it might cost me in discomfort and danger." He added that he was too poor to manage an expedition on his own and asked for three hundred dollars to support the effort—money to be spent on ponies, a wagon, a driver, and a cook.

"I like the style of your letter," Cope replied. Enclosed in his reply was a draft note for the requested money.

Cope's support for the young fossil hunter came as a surprise. The two men had never met and had no prior contact, but based on a single letter, Cope chose to invest in the up-and-coming fossil hunter. While Sternberg was not particularly interested in taking sides in Cope and Marsh's Bone Wars, he couldn't help but feel partial toward the man who had backed him.

"That letter bound me to Cope for four long years and enabled me to endure immeasurable hardships and privations in the barren fossil fields of the West," Sternberg wrote, "and it has always been one of the joys of my life to have known intimately in field and shop the greatest naturalist America has produced."

Perhaps thirty-five-year-old Cope saw himself reflected in twenty-five-year-old Sternberg's passion for paleontology. And maybe their shared lack of formal training brought them closer. In the summer of 1876, Sternberg accompanied Cope on a fossil-finding expedition to Montana Territory's Judith River Formation. The journey west was exhausting. "We traveled ten miles an hour, day and night, stopping only for meals," Sternberg recalled. Yet such hardships solidified Sternberg's friendship with Cope.

On one occasion, while stopped for some much-needed rest, Sternberg cradled Cope's head in his arms so that the famed professor might "get a few hours' rest."

They passed the time in conversation atop their horses, Sternberg acting as an attentive student while Cope regaled him with tales.

But it wasn't all head-cradling and chitchat. A potentially dire threat awaited them in Montana Territory: the newly emboldened Sioux warriors.

While at a hotel in Helena earlier that summer, Cope and Sternberg received word of Lieutenant Colonel Custer's death and defeat at the Battle of Little Big Horn (also known as the Battle of Greasy Grass). Such reports might've persuaded most scientific expeditions to turn back, but not Cope, who calculated that the circumstances created the perfect conditions to forge ahead. The Sioux warriors would be too engaged with U.S. forces to bother with a small band of fossil hunters.

Cope's gamble mostly paid off, though in August, while in the Judith River badlands, Cope, Sternberg, and the third member of their expedition, Mr. Isaac, encountered a man and woman from an unnamed Indigenous tribe rushing toward them. Mr. Isaac reached for his rifle, though upon hearing the Indigenous man shout, "Me good Indian! Me good Indian!" he stopped short of firing.

The fossil hunters had been mistaken for whiskey dealers, though once the truth was sorted out, Cope invited the pair to encamp with them for the night and to invite others from their tribe to join them for breakfast in the morning.

It made for a tense night of little sleep. But at dawn, just as Cope was washing his false teeth, six chieftains entered the camp, having accepted the invitation.

First impressions count, as Cope was reminded that day when he casually slipped his false teeth back into his mouth. The chieftains were astonished by what they'd witnessed—a man whose teeth could be inserted and removed at will.

"Do it again! Do it again!" the men shouted. Cope, emboldened by his parlor trick, was happy to oblige, and the chieftains appeared duly impressed.

"We never knew whether this hospitality was of any benefit to us, as the whole tribe went on their buffalo hunt, and we saw no more of them," Sternberg wrote, "but very likely their chiefs forbade petty stealing from our camp, for we lost nothing."

*

One hundred forty-seven years later, Tom's expedition team—including Beth, The Nephews, Ellie, and me—pile into his pickup and drive forty-five minutes outside Roundup to a private ranch where we have permission to dig. As we drive, I can't help but think of Cope and Sternberg all those years before, pulling from the earth one dinosaur fossil after the next, just a hundred or so miles from where we are now.

Tom parks his truck on a hilltop in a place he's dubbed Prairie Dog City. We've got an afternoon of digging ahead, so we carb up on sun-warmed peanut butter sandwiches. A few dozen yards away, the citizens of Prairie Dog City greet us with their prairie dog welcome—squeaking and chittering a message that roughly translates to "Get the hell off our property."

We do the next best thing: granting them a wide berth.

"Got your dig kits?" Beth asks. The kids and I nod.

"Then move out," Tom says, taking the last crisp bite of his apple.

Today, we'll be digging in the fossil-rich Lance Formation. This is land worth betting on, composed mostly of sandstone, mudstone, and clay. Of all the fossil hotspots in Montana, Lance, Judith River, and the Hell Creek Formations are where dinosaur fossil hunters like us most want to be.

We shuffle down a sagebrush slope toward a site where Tom's had some previous luck.

Upon our arrival, Tom settles into a hillside and opens his dig kit.

"Today," he announces, "we are searching for Polly."

"Who's Polly?" Ellie asks.

"Polly is a *Triceratops*," Tom says. "And she's got a sharp beak, which is why I named her Polly. Like 'Polly want a cracker'?" Tom squawks.

Ellie laughs.

"I've found portions of her frill all over here," Tom continues, sweeping his hand toward the grass-tufted landscape. "Maybe today we'll find a little more."

Hope remains high on the dig site when the day is fresh and new. Each of us takes our place in the crumbling dirt, prodding at the earth with our screwdrivers until enough ground gives way for us to gather it up in our arms. We pull the dirt forward and downward, letting it fall from the shelf to reveal what's buried beneath.

Which, admittedly, is not much. Dirt clods give way to finer dirt, and still we keep digging. Over the next hour, wishful thinking gets the better of us as we grow more desperate for fossils. Ellie and I fill our kit lids with what we believe to be dinosaur material. Each prospective fossil undergoes a gamut of lick tests and studies in coloration, and when we're still not sure, we walk it over to Tom for an ID.

"Rock, rock, rock, another beautiful rock," Tom says, unceremoniously tossing our finds back to the ground. But then he pauses, eyeing the next piece he's lifted from the kit lid. "Now this looks like a little piece of frill," he tells Ellie. "Maybe hold onto this one."

Happily, she places it in her "save" pile.

Our forearms function as bulldozers as we dig through one layer of dirt after another. We tell ourselves, again and again, that the next big find is just a brush stroke away.

I'd like to report that we unearthed Polly that day, but we didn't. Instead, all we unearthed was more earth.

We close our kits midafternoon, then follow Tom toward a nearby hill.

"Come on," he says. "I want to show you something."

We sidestep down a hillside, cutting through the dried-up stream until we reach the base of the hill.

"Anyone know what this is?" Tom asks, nodding to a two-foot-tall rock formation.

"Petrified stump?" I try.

"It's a compacted sandstone rock," Tom says. "And you know what this is inside it?"

We shake our heads no.

"*T. rex* leg," he says, nodding to a fossil with a honey-comb-like center nestled into the rock. "You can tell because *T. rex* is one of the few dinosaurs whose bones are hollow."

Stillness shrouds the landscape as even the prairie dogs have grown quiet. Kneeling, Ellie places her palms on the fossil as if granting it absolution. We have entered hallowed ground, a sacred place, and we pay our respects to the king.

The fossil upon which Ellie now rests her hand once belonged to the greatest predator of the Late Cretaceous. It transported thirteen thousand pounds of flesh, bone, and body through the forests and braided streams that once populated this landscape.

"So . . . we're just going to leave it here?" I ask.

"It would take some serious equipment to properly remove it," Tom explains. "And like I always say, it's been sitting here for sixty-six million years so . . ."

". . . it's not going anywhere," I say, finishing his line. Tom nods approvingly.

We walk a wide berth around Prairie Dog City before returning to the safety of the truck.

"Sorry, El," I say as we buckle. "I guess we didn't find Polly today."

"Yeah," she sighs, settling for the consolation prize. "Only that *T. rex* leg."

*

Some days you uncover dinosaurs; other days you're simply grateful to escape with your life. The latter was all Charles Sternberg could hope for one summer afternoon in 1876 when, at Cope's request, he climbed a ledge on a remote ridge in search of dinosaur bones.

Finding nothing, he walked toward a gorge with a thousand-foot drop. The view was beautiful and terrifying, "a scene of desolation," he later wrote, "such as no pen can picture."

Suddenly, the ground beneath Sternberg's feet gave way as he attempted to cross from one ledge to the next. Instinctively, he reached for his pickax.

"God grant that I may never again feel such horror as I felt then, when the pick, upon which I depended for safety, rebounded as if it had been polished steel, as useless in my hands as a bit of straw." He struck the rock face again and again without success, as his body slid "with ever-increasing rapidity toward the edge of the abyss."

Sternberg's life flashed before his eyes during this life-or-death moment. "Everything . . . that I had ever done or thought spread itself out before my mind's eye as vividly as the wonderful panorama of the cliffs and canyons upon which I had been gazing a few moments before." He saw glimpses of his childhood, acquaintances, and beloved mother—whose face crystallized in his mind's eye, even as he wondered "what she would think when she heard that I had been dashed to pieces."

Something spared Sternberg that day, though he could never identify what. Like Tom Hebert at his kitchen table with his gun in 2002, the world seemed to grant him a reprieve.

Sternberg's body stopped its slide on the cusp of the ledge, where he lay for an hour before regaining enough strength and composure to return to camp.

It was just one of many setbacks that summer, none serious enough to nudge the men back to the comforts of their homes. Cope, Sternberg, Mr. Isaac, and their cook endured burning days and chilly nights while pursuing Montana Territory's dinosaur fossils. Sternberg reported that water was scarce, and the food all but "indigestible." Many nights, Sternberg awoke to Cope screaming; the professor was tormented by nightmares. "Every animal of which we had found traces during the day played with him at night, tossing him into the air, kicking him, trampling upon him," Sternberg recalled. Sternberg would wake his friend, momentarily freeing him from the nightmare, which would often begin again.

I've long wondered what it must've been like to try to sleep during those dark nights in the West while the American Indian Wars raged not far from their camp. How, despite challenging conditions and various threats, the loyal Sternberg lay at Cope's side, tending to his friend and colleague's nightmares.

Indeed, there were easier ways to make a living. (Occupations, for instance, that didn't involve clinging to the high edge of a ledge). However, few jobs were more adventurous or fulfilling than discovering dinosaurs—at least for scientific-minded men like Sternberg and Cope.

*

Late in the afternoon—after gorging on the best gas station breakfast burritos this side of the Rockies—Tom, Beth, The Nephews, Ellie, and I pile into the pickup and head toward our campsite. After two nights at the ranch's guest house, Tom thought we ought to experience the total darkness of the Montana sky.

"There's nothing like it," he promised. "Just wait until you see it."

I wasn't sure how one "sees" darkness, but I was excited to try. And I was equally excited to spend the next eighteen hours completely cut off from the rest of the world. (A scary proposition for a guy who grows nervous when his cell phone loses its full signal bar.)

"Hold onto your hats," Tom says, turning right off the main road onto what I will politely call a "ranch road." But within a quarter mile—as our bodies jostle like ragdolls to the beat of the rugged terrain—any last vestige of "road" runs out. Now, we are at the mercy of grass as tall as the truck windows.

Aside from an abandoned ranch and a line of electric poles, no signs of human life exist here. For miles in every direction, all I see are grasslands that ebb and flow like a tide. On the horizon, a hundred or so cattle bow their heads—not in prayer, but to partake in the earthly delights.

Our tents are staked to the flattest land near the top of the hill. To our right, a ridge of boulders gives rise to several trees. Higher on the hill is a geological formation called Big Wall—a hundred-foot-tall sedimentary rock composed of compacted sandstone. Seventy million years ago, it was likely a terrestrial landscape where dinosaurs roamed; today, it's the best place to take in the view.

Once the tents are secure and our sleeping bags unfurled, we pause to take in that view ourselves. It is so wondrous that we dedicate the next six hours to doing little else. Seated in our camp chairs, we peer down the valley toward Roundup, fourteen miles away.

"Now this is the life," Tom sighs, his hands resting loose on his lap. It's the happiest I've ever seen him.

"Um . . . does anyone see the air mattress pump?" Beth calls.

Tom frowns. It is the grumpiest I've ever seen him.

"I told The Nephews to grab it," Tom says, turning their way. "Right, fellas?"

The Nephews' faces redden.

"I swear we packed it!" calls the older of the pair.

"We did!" cries the younger.

It seems more a matter of semantics whether it's a "missing" pump or a "never packed" pump. Either way, the ground is hard.

By early evening, a light drizzle approaches our site. We can see it coming for miles, a line of dark clouds quietly rolling in from the west. Our makeshift firepit (a metal magazine rack tipped downward to hold the flame) crackles, sending a breath of warmth our way. But it's barely enough. As the rain splatters against my forearms, I kick myself for not bringing a sweatshirt.

"Never fear," Tom says, pulling from his pickup the event tent we'd previously used for tabling at the Roundup Independence Day Extravaganza. We leap from our seats to lift its metal legs, positioning the tent high above the flame. Then we tie down the guylines—anything to prevent the next gust of wind from turning the tent into a sail.

"Who's hungry?" Beth asks. Five hands shoot high in the air.

She opens the cooler, revealing foil-wrapped packets of fish—tuna steaks, mahimahi, and salmon. We've got mixed veggies, too, which we place atop the grate beside the fish.

"Who packed the plates?" Beth asks. The answer, of course, is none of us.

Tom scans our campsite, his eyes falling to a stack of one-foot-by-one-foot boards intended for firewood.

"I got plates," he says proudly, lifting a board.

If you've never had a campfire-cooked tuna steak served on a board in the middle of Montana, you must. It's a four-star dinner with a five-star view.

FIG. 11. Ellie enjoys a campfire-cooked tuna steak on a board on private land in Musselshell County, Montana. Author photo.

Ellie and I shovel fish into our mouths at a rate that can only be described as rude. Our manners have gone missing with the air mattress pump.

"Good?" I ask Ellie.

"Great," she grins, returning to the truck bed for a second helping of vegetables.

At meal's end, we "do the dishes" by tossing the butter-soaked boards into the fire. The flames rise hungrily.

In the distance, the sky changes from bluish-black to black and back again. Clouds enter the scene and depart like background actors. But at 8:40 p.m., the threatening storm recedes, leaving an orange hue lingering above the tree line to the west.

"There's only one place to catch a sunset like this," Tom says, hustling uphill toward Big Wall.

We follow; each *swish-swish* of the tall grass brushing our legs sounds like whispers from rattlesnakes. Indeed, rattlesnakes are all I can think about as Ellie crashes through the grass behind The Nephews.

I do a little worst-case-scenario math:

If a child were to be bitten by a snake fifteen miles from town, how long would it take before we could get a dose of antivenom? The answer—if I'm even remotely correct—does little to ease my anxiety. How quickly one of the most beautiful moments of my life could turn into the ugliest.

But on this night, we are spared any snakes. We hike higher up a rocky embankment until we reach Big Wall. From the top of the sandstone cliff, we see nothing but grasslands and winding streams. Perched on the precipice, we all but lose our breath at the sight of it. Ellie tiptoes one foot closer to the edge, then another, until I beg her to return to me.

"Dad, I'm fine," she assures me. "I'm not going to fall."

(I imagine Sternberg felt similarly before he ended up clinging to the gorge's edge.)

"You've got the surefootedness of a mountain goat," I assure her. "Now please, do Daddy a favor and back away from the ledge."

Sighing, she backtracks, granting me peace of mind.

I've never seen our country quite like this—from this height and vantage point on the eve of the Fourth of July. What we see before and beneath us resembles an Albert Bierstadt painting—a perfect palette of colors splashed with lavish bursts of light. The natural splendor is nearly gratuitous—the greenery hemmed along the streams below, paired with the fiery orange from the sun.

Tom climbs to the top of a cliff, lifting his arms above his head. He is as happy as Sternberg after not falling to his death, as happy as Cope after finding a new fossil. Tom lets out a whoop, which echoes through the ages, bouncing from Big Wall to the America we see far below.

I pray for a few more minutes of Albert Bierstadt's light. For one more glimpse of Ellie's wide-eyed wonder. One more second of her carefree smile spread across her face.

No matter what preadolescent pain she has waiting for her back home, nothing can touch us out here.

A few minutes past 9:00 p.m., the last glint of purpled sun descends beneath the horizon. The afterglow is the only light guiding us back toward the tents. Darkness comes with a vengeance, along with a stiff wind and the next spritz of rain.

While I worry about rattlesnakes, a nasty swarm of biting flies prepares for their sneak attack. They appear out of thin air, a squadron of tiny-teethed bugs nibbling onto our flesh.

"Ahhh!" Ellie cries, taking off at top speed. "They're everywheeeeeeere!"

The flies zip like dogfighters, divebombing her ears and eyes. I cover her as best I can, absorbing a few fly bites myself.

When we reach the tent, Beth comes to our rescue.

"Here," she says, plucking a nearby plant, "rub this on."

"What is it?" Ellie asks.

"White sage," Beth says. "It's an old Indian trick. The smell will keep them away."

No one needs to tell Ellie twice. She rubs the plant across her face, and the flies, indeed, retreat.

Now that we've avoided one inconvenience, we're faced with another: the next round of rain begins almost immediately.

"Well, that's our cue," I sigh, rising from the camp chair. "I think we've had about all the adventure we can handle. Night, everyone."

Everyone echoes a reply.

Tom, Beth, and The Nephews huddle around the fire for a few more minutes while Ellie and I remove our shoes—hopping on one foot and then the other—before falling into the lip of the unzipped tent.

Tucked in our sleeping bags, we begin our nightly ritual—a recap of the day's highs and lows.

"My high," Ellie says, "was probably sitting around the campfire or digging for dinosaurs. Oh, and the sunset. Oh! And when we ate the fish off the boards."

"So . . . pretty much the whole day?" I laugh.

"Yes," she nods solemnly, "the whole day was my high for the day."

"And the low?" I ask.

"Those stupid flies," she grumbles.

"Think of it this way," I say, "at least now the tent smells like sage."

"Yeah, yeah," she yawns, closing her eyes and pulling her sleeping bag tight. "Good night, Dad."

"Night, hon."

We fall asleep to the sound of tapping rain.

*

In all, Sternberg served as Cope's assistant for eight seasons. They didn't always agree, though they cared deeply for one another.

In mid-April 1897, while digging on Cope's behalf in Texas, Sternberg received a letter informing him of the professor's death.

"I had lost friends before and had known what it was to bury my own dead, even my firstborn son, but I had never sorrowed more deeply than I did now over the news that in the very prime of life, in the noonday of his glorious intellectual achievements . . . the greatest naturalist in America had passed away with his work undone."

FIG. 12. Paleontologist Charles H. Sternberg poses on a porch in 1931–32. Sternberg, George Fryer 1883–1969, "066_02: Charles H. Sternberg" (2021). *George Sternberg Album #5 - Scrapbook 1931–1932*. 165, Forsyth Library Special Collections, Fort Hays State University.

Sternberg continued the work for decades, though while visiting one of his dinosaur discoveries later in life, the devout Sternberg felt compelled to pen the following:

"My own body will crumble in dust, my soul return to God who gave it, but the works of His hands, those animals of other days, will give joy and pleasure 'to generations yet unborn.'"

Like Cope, Sternberg dreamed that his fossil-finding legacy might extend beyond his life.

It did.

*

I woke, shivering, sometime around 3:30 a.m. Despite my many reminders for Ellie to pack the appropriate sleepwear, I failed to do so myself. My T-shirt is damp from the rain, and my long underwear is back at the ranch.

Outside the tent, I see a spark from the revived fire, followed by the crackling sound of wet wood drying. In my sleepy state, it's unclear if this is the beginning of an unplanned prairie fire or something a little more controlled. Judging by Tom's silhouette, I suspect the latter.

I unzip the tent flap and shiver my way toward the flame.

"Early bird gets the worm," I say.

Tom chuckles, staring at the sky.

"That lack of air mattress really hurt us," he says, shifting uncomfortably in his camp chair. "Couldn't sleep a wink."

"Yeah, well, our sleeping pads weren't that helpful, either," I commiserate.

I excuse myself to pee behind some boulders, hidden by darkness as thick as velvet offset only by the stars. When I return to camp, I pause at the sight of Tom's silhouette sunk low in his chair. His profile is only occasionally visible, depending on the size of the flame. He appears deep in thought, and I suspect I know about what.

Shortly before bed, he'd vented to me about his frustrations with the paleontology community. Despite thirteen years of experience digging dinosaurs, most of the community wouldn't consider him a particularly valued member of their field.

"All because I don't have the right piece of paper with my name on it," he'd sighed. (Though his critics might note that Tom can be challenging to work with too).

A few dozen feet ahead, Tom settles into the predawn darkness, taking his hands out of his pockets to warm them by the flame.

I want to snag the seat beside him, return to my "professor mode," and offer unsolicited advice. But what do I know about the world of paleontology? Or the recriminations tossed back and forth among professional paleontologists, amateur paleontologists, and commercial fossil hunters, in which Tom now finds himself caught somewhere in the middle?

I might've kept Tom company until dawn, but I feared I'd say the wrong thing. Employ the old company line urging him to return to school, finish his degree, and earn that piece of paper and alphabet soup. It worked for me. Why wouldn't it work for him?

But it's not my place. Not here, not now.

Instead, I say, "Good night, my man. I'll see you in the morning."

*

By 6:00 a.m., the tent is folded in its carrying case, and our sleeping bags are compressed in their sacks. Shivering, Ellie wraps herself in her green dinosaur blanket, tilting her head westward for one last look at Big Wall.

"You mind driving us back to the van?" I ask Tom. "I guess we ought to be hitting the road."

"Not at all," Tom says, even though "driving us back to the van" entails off-roading for a mile or two before even getting to something resembling a road.

"Want to know something funny?" Tom says, reaching for his truck keys. "We just found the air pump."

I try—and fail—to stifle a laugh.

"Well," I smile, "at least The Nephews are off the hook."

Beth emerges from the tent, droopy-eyed.

"Did you hear we found the pump?" she asks miserably.

I nod. She sighs.

"Thank you two so much for joining us out here," Beth says, hugging Ellie and me. "We loved sharing this part of the world with you."

"Thank you," I say. "Eating fish off wooden boards was the best meal of my life."

"You know," she muses, "it *was* pretty good."

Ellie and I hop into the truck as Tom begins the long, slow drive out of the country. Halfway down the rolling hills, we spot a deer leaping high above the grass. Moments later, a coyote—scared by the truck, no doubt—takes off in the deer's direction.

"I was checking the weather," Tom says as we bump and thump toward the road. "And it's snowing somewhere in Montana right now."

I shake my head.

"I mean, can you even believe that?" he asks incredulously. "Snowing on the Fourth of July?"

I smile at the phrase, and I can already hear the tune.

*Why does everything out here sound like a country and western song?*

FIG. 13. Ellie takes the Junior Ranger oath at the Lewis and Clark Interpretive Center in Great Falls, Montana. Author photo.

# 6
# Of Mammoths and Men

**JULY 4–5**
*Great Falls, MT*

Somewhere between Lewistown and Great Falls—as the clouds hang like shawls in the sky—the van's touch screen prompts an eerily prescient message: *Would you like to take a break?*

I lift an eyebrow toward the ghost in the machine.

Would *I* like to take a break? From what? Driving? Modern society? The existential crisis of a planet on the brink of collapse?

Yes, I believe I would. However, I have no idea how to relay my response to the touch screen. Do I press a button? Offer some verbal acknowledgment to my robot overlords?

"Yes," I say aloud to no one. "I would like to take a break."

Now Ellie's the one lifting an eyebrow.

"Dad," she says, "are you . . . okay?"

"Are any of us?" I wonder aloud.

"Well, I'm good," she says cheerily, offering a thumbs-up.

I shift my sleepy gaze back to the touchscreen, where all traces of the message have vanished.

"Then I'm good, too," I yawn. "Daddy just needs coffee."

*

On July 4, 1805—exactly 218 years before our early morning arrival in Great Falls—Captain Meriwether Lewis, Second Lieutenant William Clark, and their Corps of Discovery inhabited the same terrain. They, too, were likely in need of coffee.

President Jefferson had dispatched them the previous year, hoping their expedition might establish trade routes, halt any potential encroachments by European powers, and foster relationships with Indigenous tribes in the newly acquired lands of the Louisiana Purchase.

In addition to these economic and geopolitical objectives, Jefferson also asked the expedition to serve as amateur scientists, reporting on the West's geography, geology, zoology, and botany.

Jefferson also encouraged Lewis and Clark to look for ancient fauna, most notably mastodons, which he often called mammoths, and which he believed might still be alive and well and roaming the American West.

Such a hypothesis seems like lunacy today, but it wasn't during Jefferson's time.

Believing it unlikely that nature could exterminate entire species, Jefferson held firm to his conviction that species whose fossils were found might still exist. Had Jefferson come into possession of dinosaur fossils, one wonders if he might have imagined dinosaurs, too, lumbering across the Great Plains. (Though there were no dinosaurs known during Jefferson's life, English biologist Sir Richard Owen first coined the term *Dinosauria* in 1842, sixteen years after his death.)

While Jefferson's passion for science and natural history hardly subsided during his White House years, his time for scholarly inquiry and exploration was limited. I like to think a part of him lived vicariously through his agents, who, for 862 days, explored the inner regions of the western half of the country, which Jefferson himself would never see.

From grizzly bears to prairie dogs, no fauna was too big or small for Lewis, Clark, or the members of their party to take note. And what they lacked in mastodons, they made up for with documentation on hundreds of other animals and plants.

"What a field for a [botanist] and a [naturalist]," Clark wrote dreamily of the country.

But on July 4, 1805, the corps discovered something even more wondrous than a new species: a place to put in their boats beyond the waterfalls for which the city of Great Falls was later named. After weeks of portaging, they could at last return to the river.

Thus, two events were worth celebrating on that Fourth of July: the country's founding and the end of portaging.

In his journal that day, Lewis described the evening's festivities.

"We gave the men a drink of [spirits], it being the last of our stock, and some of them appeared a little sensible of it's effects," Lewis wrote. "The fiddle was plied and they danced very merrily until nine in the evening when a heavy shower of rain put an end to that part of the amusement."

For dinner, the crew enjoyed bacon, beans, and buffalo beef.

"In short," Lewis concluded, "we had no just cause to covet the sumptuous feasts of our countrymen on this day."

*

All these years later, Ellie and I covet their feast. Hungry and tired (our last meal was the previous night's board-plated fish), I'd have traded most of my earthly possessions for a generous serving of whatever buffalo beef is.

While stopped at the red light in Great Falls, I spot a sign for the Lewis and Clark Interpretive Center. I make a mental note to visit, right after we attend to more pressing matters: a van desperately in need of an oil change.

We park the van at the side entrance of a big-box store near the oil change bays. I'm so tired that when I relay the van's make, model, and year to the technician, I only manage two out of three correctly.

After getting the year right on the second try, the technician types my information into his computer and then glances up worriedly.

"Uh-oh," he says.

"*Uh-oh* what?" I ask.

"Looks like we're fresh out of oil."

I look around the cavernous oil and lube department attached to the even more cavernous big-box store.

"So . . . no oil . . . anywhere?" I ask, bleary-eyed.

"Dude, I'm joking," the technician grins, directing me toward row upon row of the stuff.

But I'm so tired that his joke hardly registers.

The technician gives me the once over.

"Why don't you get a little rest?" he says, relieving me of my keys. "It might do you some good."

I don't always take self-care tips from touch screens and oil change technicians, but on this day, I do.

Ellie and I wander through the big-box store until we arrive at the bathrooms, which, considering our recent no-frills accommodations, suddenly resemble an all-inclusive resort.

"Okay," I say, "time for our sink showers."

"What's a sink shower?" Ellie asks.

"What do you think?"

Upon entering our respective bathrooms, a glance in the mirror reveals what may have tipped off the technician to my need for rest. In the long history of bedraggled middle-aged men, I am the most bedraggled of all. True, I haven't shaved or showered in days, but what's most unnerving are the bags beneath my eyes, big enough to carry a couple of baby joeys.

I splash a gallon of cold water on my face in search of the person underneath. After some scrubbing with a paper towel, he starts to emerge, only slightly worse for wear.

Five minutes later, Ellie and I rendezvous near the bathroom water fountain. She's drenched.

"Whoa," I laugh, "I didn't mean a literal shower."

"I know," she says, wringing the water from her hair. "I mean, I know that now . . ."

"Were you in the splash zone or something?"

"Da-ad . . ." she says.

To kill time, my dripping daughter and I load up on provisions—apples, oranges, potato chips, and several bottles of huckleberry lemonade in honor of Tom Hebert.

Our provisions secured, we return to the oil and lube department.

"How'd it go?" I ask the technician.

"Well, the oil change went fine, but we noticed your windshield wipers are junk," he says. "You've been scraping bare blades against the glass for some time now."

"Oh," I say. "Huh."

As travel setbacks go, bad blades sure beat eighteen miles of portaging. More concerning, though, is that I hadn't even noticed.

*

When Lewis and Clark had problems, they noticed. One involved making the expedition-imperiling decision between which of two rivers to follow: the muddy waters of the north branch (which Lewis named the Marias River) or the clear waters of the south (the Missouri River). The right choice would keep them moving steadily toward the Columbia River; the wrong choice could add hundreds of miles to their route. For five days, the expedition explored both branches before Lewis and Clark—against the judgment of their party—selected the south branch.

"[The men] said very cheerfully that they were ready to follow us any wher[e] we thought proper to direct," Lewis wrote on June 9, 1805, "but that they still thought that the other was the river."

In this instance, Lewis and Clark were correct. While their float down the Missouri River yielded no mastodons, they made many other discoveries.

Many of which we learn of during our visit to the Lewis and Clark Interpretive Center, twenty-five thousand square feet of exhibitions and art along the banks of the Missouri River. As American rivers go, the Missouri—a tributary of the Mississippi, and the longest in the nation—has earned its bragging rights, including safely transporting Lewis and Clark and their Corps of Discovery from its mouth in St. Louis to its headwaters near Three Forks, Montana.

The story of a river is the story of the people who travel it. For thousands of years, the Missouri River (known as Peki-tan-oui on early French maps) has been a watery thoroughfare for food and trade throughout the Great Plains. Originating in the Rocky Mountains, it flows east and south for more than 2,300 miles, winding through seven states and serving as a boundary for many of them—Kansas and Missouri, Nebraska and Missouri, Nebraska and Iowa, and South Dakota and Nebraska—before reaching Bismarck, where we'd first glimpsed it days prior, then onto St. Louis, where it meets with the Mississippi. When traveling downstream, it's a blessing. Paddling upstream (and contending with five separate sets of waterfalls) can be a curse, as Lewis and Clark learned the hard way.

But on this sun-drenched Independence Day, as we peer out through the interpretive center's wall of windows, all we see is the river's beauty. It is not an obstacle to overcome but a paradise for the white pelicans, who preen themselves in its slow-churning current.

What Great Falls lacks in dinosaurs, it makes up for with Lewis and Clark—whom the city has fully embraced, even though the explorers probably spent most of their time cursing the region's unnavigable falls.

Taking a "when in Rome" approach, Ellie and I double down on Lewis and Clark. For the next hour, Ellie completes the interpretive center's Junior Ranger Activity Journal—fourteen pages of mazes, code-breaking, and fill-in-the-blanks, with a page reserved for personal reflection. In the bubble that asks for her thoughts, Ellie writes, "I feel great when I hear that people helped Lewis and Clark, especially Sacagawea. When I hear about how hard Lewis and Clark worked, I feel so proud of them."

Indeed, it makes for an inspiring story—how, over twenty-eight months, Lewis and Clark traveled eight thousand miles with their crew, mapping and "discovering" a landscape previously and primarily known by Indigenous tribes, including the Osage, Sioux, Cheyenne, Crow, Shoshone, Hidatsa (also called Minitari), Blackfeet, Chinook, and Mandan, to name a few. On Jefferson's orders, Lewis and Clark came not as conquerors but as goodwill ambassadors, bestowing tribal leaders with peace medals. But the goodwill would not last—not when waves of white settlers began taking up residence on the land upon which they lived. For Lewis and Clark, the pain they caused resulted from the path they forged. In part because of their "discoveries," destruction followed.

After completing the Junior Ranger Activity Journal, Ellie returns to the woman at the front desk to claim her prize. The woman pages through the journal, and after confirming that all mazes have been navigated, all codes broken, and all blanks filled, she says, "Are you ready to take the Junior Ranger oath?"

Ellie looks perplexed; she was mostly just in it for the free sticker. But she does the most American thing imaginable on this Fourth of July. On the woman's instructions, she poses before an American flag and a bust of Thomas Jefferson and, with her right hand raised, repeats the Junior Ranger's sacred oath—which includes showing respect for Native homelands, studying history from multiple perspectives, and playing safely in nature.

"Congratulations, Junior Ranger," the woman says, handing her a ranger badge sticker.

Ellie smiles—that's more like it.

*

Shortly before sundown, we pack sandwiches and chips and search for a picnic spot at nearby Rainbow Falls. Since Lewis and Clark managed 8,000 miles with only a few early nineteenth-century navigational tools, I'm confident I can find a 47-foot-high and 1,300-foot-wide waterfall with the help of my GPS.

"Are you sure we're going the right way?" Ellie asks ten minutes later, as we wind along a river road, no hints of waterfalls anywhere.

Sighing, I pull over to refer to the GPS.

"Maybe we should . . . turn back," Ellie suggests.

"Honey," I say, facing her, "did Lewis and Clark 'turn back' when things got tough?"

"They portaged," she reminds me. "For eighteen miles!"

"Thanks a lot, Junior Ranger," I grumble.

Had we continued for another half mile, we'd have managed a view of the falls, though not a very good one—half hidden behind the infrastructure for the hydroelectric dam. We'll marvel at the true wonder of the falls tomorrow. But for now, there are sandwiches to eat.

We turn back, retreating to the south side of the river, where we spot a pull-off on the River's Edge Trail. We set up dinner at a picnic table alongside a refurbished train caboose. We unroll the vinegar-slick wrappers from our sub sandwiches, then lift the upper half of the bread to sprinkle crushed chips atop the salami and shredded lettuce.

All those years before (and just a few miles away), Lewis, Clark, and their crew had enjoyed their "sumptuous feast" to celebrate the Fourth of July. We celebrate ours in a similar fashion while enjoying a similar view.

"*Bridge to Terabithia*?" I ask, reaching for the paperback. Ellie nods as half a head of shredded lettuce falls from her bulging cheeks.

I flip to our dog-eared page, reading aloud from the story of two friends who take refuge from their real-world problems by creating an imaginary world just for them. But beyond that magic is better magic—a tale of a burgeoning friendship between two kids with so much to share. Ellie stares dreamily toward the river as I read of Jesse and Leslie's first meeting, how Leslie approaches her new neighbor and suggests they become friends. Just like that, the deal is done.

Ellie smiles at the simplicity of their friendship on the page. How desperately she wishes it were that easy in life.

When Meredith and I watch our smiling daughter sling her backpack over her shoulders and embark upon another day at school, what can we do but try to match her brave smile with our own? Tell her what a great day she's got ahead of her—we hope?

Beyond the safety of our front door, each of our family members travels a more uncertain path. We never know what lies ahead for us, nor which fork in the road—or river—we ought to take.

I glance up from the pages of *Bridge to Terabithia* to catch Ellie peering out at the mighty Missouri just ahead.

"I love this book," Ellie whispers.

Indeed, it's a nice place to escape to for a while.

*

Although Lewis and Clark fell short in their search for living mastodons, they may have done more for dinosaurs than was initially thought. On

July 25, 1806—a year removed from their portaging troubles near Great Falls—William Clark discovered a large fossil near Pompeys Pillar, just east of modern-day Billings along the Yellowstone River. In his journal, Clark described it as "the rib of a fish which was [cemented] within the face of the rock." It made for a strange fishbone, measuring three inches in circumference and approximately three feet long. Later, scientists determined that the fossil possibly belonged to a *Hadrosaur*, not entirely unlike the fossil Ellie had discovered.

Indigenous people have discovered ancient fossils for thousands of years. However, according to some paleontologists, Clark's find may be the first documented dinosaur fossil discovery in North America.

As if Lewis and Clark hadn't discovered enough, some now add "dinosaur fossil" to the list.

*

As the sun goes down, so, too, do the temperatures. Back home in Wisconsin, we'd grown used to a nightly drop of ten or so degrees, but here in Montana, the drops are a little more precipitous. If the weather app is to be believed, we're looking at a twenty-degree plunge in temperature on the Fourth of July. Even worse, the high for the following night is expected to be in the forties.

This is all I needed to know to ditch the tent and upgrade to the Great Falls KOA one-room cabin, which I arranged moments after we arrived in town. It has everything we could dream of—beds, a desk, even a heater. Its log cabin decor provides the ambiance of the outdoors while sparing us the teeth-chattering.

After our riverside picnic, Ellie and I return to our cabin's porch swing, where we soon hear the magical and unexpected twang of guitar and banjo from somewhere beyond the trees.

Ellie's eyes widen.

"Terabithia?" I ask.

Grinning, Ellie leaps from the swing, leading us down the wooded path toward the outdoor campground kitchen. Part gazebo, part pavilion, during

FIG. 14. B.J. and Ellie upgrade their tent to a cabin at the Great Falls KOA. Author photo.

cooking hours, the outdoor kitchen serves as an ideal place for washing dishes, microwaving burritos, or boiling pots of water. But come nightfall, it's transformed into an intimate concert venue—its benches and picnic tables overflowing with twenty or so listeners. Ellie and I slide into a couple of empty seats near the back.

It's not Terabithia, but it's close.

"Hi everybody, I'm Richard," says the man with the banjo.

"And I'm Diane," says the woman with the guitar.

"And we are the Rivertown Rounders," Richard smiles. "We play country songs, folk songs, gospel songs, bluegrass songs, children's songs, train songs, Cajun songs, Irish songs, cowboy songs, dying cowboy songs, drinking cowboy songs, and hobo songs. So if you've got a song you'd like to hear, just let us know."

Richard and Diane have played the KOA camping kitchen almost every summer night for thirty years.

I repeat: *Almost every summer night. For thirty years.*

Which means they've been taking requests for thirty years. Which means they've been playing John Denver's "Take Me Home, Country Roads" almost every summer night for thirty years. Given the region's mountainous terrain, I'd expected "Rocky Mountain High" to rank high on the request list. Not so, Richard says. People want "Take Me Home, Country Roads."

Always and forever.

Indeed, we'll hear them play it tonight and tomorrow night too. They'll invite the crowd to sing along on both occasions, forming a camp choir whose voices converge into near-perfect harmony. Even Ellie, who's hearing the song for the first time, picks up the lyrics instantly. Entranced, she sings of a place she can't even find on a map. No matter that the song is about West Virginia and we're in Montana. Geography is a state of mind.

Between songs, Richard and Diane work the crowd with the seasoned professionalism of cruise ship entertainers. After thirty years of performing their schtick, I'd have thought they'd grown weary, but they haven't. Their interest in every listener is astonishing.

"Where are you two from?" Diane asks, pointing to us.

"Wisconsin," I say.

"Wisconsin," she nods, plucking her guitar. "And what brings you to Montana?"

"Dinosaurs," I say.

"Dinosaurs," she chuckles. "Well, we've got a few of those."

After the show—as Richard and Diane pack up their gear—Ellie and I wander their way to chat more about Montana, dinosaurs, and the secret to thirty years of strumming and plucking for the benefit of strangers.

"How do you do it?" I ask. "What keeps you coming back here, night after night, for all those years?"

"Well," Diane shrugs, "we just love it."

Richard nods. It's as simple and complicated as that.

"Are you two staying up for the fireworks?" Diane asks, closing her guitar case.

"Yup!" Ellie says.

"Well," I say, tempering expectations, "we're going to try."

It's been a long day, one that began sometime before 6 a.m. as Tom drove us from our remote campsite back to the ranch. That we are now in Great Falls, chatting with the Rivertown Rounders, seems impossible.

Retrieving our toothbrushes and towels from the cabin, Ellie and I make our way toward the bathrooms. Following our bedtime routine, we walk beyond the campground along a gravel lot for an unobstructed view of the sky. We are surrounded by strangers who join us in awaiting the light show to come. However, for the moment, the stars are their own light show. When viewed in the proper darkness, the constellations tell stories as big as dinosaurs—Orion, Aries, Cassiopeia.

I imagine Lewis and Clark staring up at the same stars from this place centuries ago. Could they have dreamed of the country we have become? Or how their route would one day be connected by interpretive centers, historical markers, and flora named in their honor?

Or can we time-travel back even further, say sixty-six million years, when one ancient creature or another stomped, swam, or soared across this landscape?

Some nights, the universe just feels overwhelming. I close my eyes, release a breath, and take a moment to imagine the bones beneath my feet and the stories stitched in the sky.

I'm brought back to the present by a pop that echoes across Great Falls, Montana; soon, hundreds of fireworks follow. Standing a few feet behind my daughter, I watch her silhouette, how she stands, unmoving, with her towel draped around her neck. The fireworks colors come and go, searing the sky before burning back to darkness. Ellie's half-shadowed face turns toward me well before the fireworks fizzle out. She yawns. I take my cue.

"Ready for beddy?" I whisper.

Clutching our toothbrushes and towels, we sleepwalk back to the campground. Upon reentering, we confront a pair of dueling paths: One will lead us straight back to the cabin, while the other will take us there via a slightly more circuitous route.

The stakes are high; the wrong choice could add full minutes to our journey. At least there are no waterfalls to contend with.

"Dad," Ellie asks, "which way?"

As I try to decide, I think not of Lewis and Clark but of John Denver.

I take a chance, veering to the right.

*Take me home, country road,* I think.

FIG. 15. B.J. and Ellie enjoy breakfast in Choteau, Montana, with Susan Luinstra, National Rural Education Association's 2007 Teacher of the Year. Author photo.

# 7
# Maternal Instinct

### JULY 6

*Great Falls,* MT →*Choteau,* MT→ *Bynum,* MT→ *Shelby,* MT

We wake to squeals on our second and final morning in Great Falls. We first heard rumors of the campground's petting zoo during check-in, though so far, we haven't seen it for ourselves. Newly inspired by Lewis and Clark, we strike out in search of the "zoo," following the squeals to the far side of the property, where we're greeted by a rooster, a chicken, a goose, a goat, and the squealers themselves—a mamma pig and her eight very needy piglets.

A dozen of us have gathered to witness momma pig parading around her pen. She's trailed by her litter, all of whom jockey for position along the moving target of her swinging teats. At last, she collapses into the dust for a feeding—her famished little oinkers trembling in gratitude as their suckling begins. Except for one piglet, whose small size distinguishes him as the runt. It's a scene straight from *Charlotte's Web*—the piglet who needs the most receives the least. No matter where he goes, the teat is always occupied. At last, he resigns himself to the back end of his mother. And then, in what seems a final indignity, Momma Pig releases a stream of urine as powerful as the Missouri, nearly washing her runt away.

On the far side of the fence, we humans provide our devastating commentary—"He's too weak"; "He's going to die"; "Last week, there were nine, you know."

"Come on, Momma," a woman whispers, her fingertips clinging to the fence. "And come on, little guy."

The runt shakes off the piss, gathers his bearings, and unsteadily returns to his hooves.

I want to leap the fence, push aside some heartier piglet, and slide the runt into the newly opened spot—anything to deny nature its natural course. Yet today's fence-hopping heroics won't be much help to him tomorrow.

"Dad," Ellie whispers. "Is he really going to die?"

The runt stares directly at us, awaiting my verdict.

"Well, hon," I begin, eyeing his wobble. "I guess it's hard to know . . ."

One by one, we humans excuse ourselves from the scene as the piglets make proverbial pigs of themselves. Except for our runt, who sniffs at the air like it's nourishment.

"Why didn't the mom just help him?" Ellie asks as we walk away.

It's a big question, even for a not-so-little girl.

*

We drive an hour northwest of Great Falls until entering the city of Choteau. Situated in northwestern Montana—geologically speaking, within the Two Medicine Formation—Choteau is where the mountains meet the plains. Its vast swaths of ranchland, abutted by jagged-peaked mountains, create the ideal conditions for wildlife to thrive. There's no shortage of mule deer and elk, though you needn't look too hard to spot the grizzlies and black bears too.

A "Welcome to Choteau" sign doubles as the city map, directing us to the post office, the schools, and the Old Trail Museum—our third stop on the Montana Dinosaur Trail. Since the museum doesn't open for another hour, we park across the street at the Outpost Deli, a cozy restaurant decorated with paintings of fish and flowers on its wood-paneled walls. We're here on the recommendation of Susan Luinstra, who agreed to meet us for breakfast.

A few weeks back, while researching the nearby town of Bynum (population 25), I stumbled upon a TODAY *Show* segment from 2017, in which Susan (and her unique teaching style) momentarily propelled Bynum into the national spotlight. The attention was due to Susan's morning dance lessons, a tradition dating back to the Depression era. The idea was developed by Ira Perkins, a local rancher and schoolteacher who had educated children in the area for more than fifty years. Initially, his morning dance lesson was a deterrent to keep high schoolers from smoking. (If students could foxtrot and waltz, he reasoned, fewer would sneak cigarettes during the town dances.) Perkins's plan worked and had the added benefit of serving as a great start to the students' day.

After observing the power of Perkins's dance lessons, the newly hired Susan was determined to keep the tradition alive, ensuring that Bynum's schoolchildren never missed a step.

Susan, now in her early seventies, flags us down from a table near the back.

"You must be B.J.," Susan says. "And you," she continues, "must be Ellie!"

Ellie beams at what feels like a maternal burst of love.

I'm eager to learn everything I can from this woman, whose achievements far exceed her fifteen minutes of fame. In 2007 Susan was named the National Rural Education Association's Teacher of the Year, a deserving tribute to her impressive thirty-eight years of service. Although Susan retired from Bynum School in 2019, she's always nearby. No one has forgotten her commitment to the young people in the area.

No sooner does our waitress take our order than Susan mentions that she once taught the waitress's grandson.

"He went to our school right before I retired," Susan says. "He's in high school now, and he just started rodeo. Now he rides bulls."

"Did you teach bull riding too?" I ask.

"Nope," she laughs. "I never taught that."

However, she taught almost everything else to everyone else. There was no grade she couldn't handle, no subject she couldn't teach.

But nothing inspired her more than science—and dinosaurs most of all.

"I was just in the right place at the right time," Susan says.

While the Two Medicine Formation had been known to contain fossils, paleontologists often preferred centering their searches to the Judith River Formation to the east and Oldman Formation to the north, which were believed to contain fossils of superior scientific value. But not Dave and Laurie Trexler, and Dave's mother Marion Brandvold, all of whom turned their attention to the less popular Two Medicine Formation on which they lived. Over the years, their searches would yield incredible discoveries, including the first baby dinosaur fossils found in North America, on a site soon named Egg Mountain.

In 1995—seventeen years after Marion Brandvold discovered the juvenile bones—Susan Luinstra participated in an on-site paleontology course on Egg Mountain. Sponsored by Museum of the Rockies, Susan and other regional teachers dug for dinosaur fossils throughout the week. Their days were spent digging in the dirt, and their nights were spent studying the stars.

"It was one of the most moving learning experiences of my life," Susan says. And she wanted to share it with others.

After her week on Egg Mountain, Susan persuaded one of the paleontologists to allow her students the chance to dig. Soon after, Susan and eight of her students took a field trip to Egg Mountain, where they spent the day digging *Hadrosaur* bones.

Even when the on-site experience ended, their work continued. A paleontologist entrusted Susan with leading her students in cleaning and prepping several large fossils, including a *Hadrosaur* humerus, which the students later saw displayed at Museum of the Rockies.

"Once you dig, it's always in you," Susan says, reaching for her coffee. "It drives you to learn something new."

*

Susan accompanies us across the street to the Old Trail Museum, where the tinkling bell on the door announces our arrival. Susan introduces us to forty-one-year-old Sean Doyle, the museum's director, whose first act is to stamp Ellie's dinosaur passport in the box marked *Maiasaura*—the museum's specialty dinosaur, given its proximity to where the species was discovered.

Sean leads us to a room full of fossils, drawing our attention first to a replica of Marion Brandvold's juvenile fossil finds.

"Marion showed some bones in a coffee can to Jack Horner," Susan explains. "She didn't realize it was a baby *Maiasaura*."

Alongside the juvenile replica is a re-creation of the *Maiasaura* eggs near a sculpture of a *Maiasaura* momma with her duckbilled face peering toward us.

"This was the first place in North America where they found dinosaur eggs," Sean says, "so it was already a big deal. But to find the eggs *and* the juveniles alongside the adults had the largest ramifications on paleontology because it revealed that they were herd animals."

And family-oriented herd animals at that.

A certain pride comes with being the "best place" for something, even when your place is considered off the beaten trail. As Sean confirms, Choteau, nearby Bynum, and the entirety of the Two Medicine Formation is one of the best places in the world for digging up dinosaurs.

The fossil record confirms it. Over the years, the Two Medicine Formation has yielded rare fossils from an array of dinosaurs: *Maiasaura* (the paternal ones), *Triceratops* (the three-horned ones), *Pachycephalosaurs* (the domed-headed ones), *Ankylosaurs* (the shielded tanks), *Dromaeosaurs* (raptors), and *Tyrannosaurs* (no description needed).

While these fossils have helped put Choteau and nearby Bynum on the map, they've also put them on the Montana Dinosaur Trail.

"A significant percentage of people who come to our museum are traveling the Montana Dinosaur Trail," Sean says. "And it does lend some notoriety to us. Being on the trail is kind of a big deal in this state. It really provides some prestige."

"And not everyone makes the cut," I say.

Sean nods.

"There's a big debate, even within the Dinosaur Trail, because some of the dinosaur museums are run by paleontologists. And they're kind of leaning towards the academic treatment of the subjects," Sean explains. "Other people are not trained paleontologists, but they're great fossil hunters and do a great job of getting them out of the ground safely."

"And they have good knowledge," Susan adds.

Sean nods. "But the way they go about procuring their fossils can be different in a lot of very important ways."

While professional paleontologists meticulously collect data on both the site and the specimen, the removal of the fossil itself is foremost on the minds of many amateur paleontologists. This is not to say that amateur paleontologists are incapable of recording scientific data, just that some don't—at least not as scientifically as paleontologists who've obtained the necessary "alphabet soup" after their names. This prompts the ire of many academically minded paleontologists who've dedicated their lives and livelihoods to prioritizing data first.

Montana Dinosaur Trail members are hardly in open conflict on the matter. ("Nobody's yelling at each other," Sean assures.) But he also acknowledges what many others have privately shared with me: MDT members have varying perspectives on how best to dig, preserve, and display dinosaur fossils to ensure that science, ethics, and tourism are all considered. Given these differences, the Montana Dinosaur Trail occasionally seems like an ill-fitting alliance. They are bound together by geography, dinosaur fossils, and passport stamps. They are divided by vast differences in funding, staffing, and scientific support.

Nowhere is this more apparent than in the differences between Museum of the Rockies and every other stop along the trail. In 2022 Museum of the Rockies brought in over $9.6 million in revenue, more than all the other MDT stops combined. Their assets top $34 million. According to a 2014 report, Museum of the Rockies—the state's most visited museum—added $48 million to Bozeman's local economy. Regarding revenue, Museum of the Rockies far surpasses its MDT partners.

I am hardly begrudging Museum of the Rockies its success. Theirs is a model that works. When grants, private donors, and membership revenue support museums, museums support local economies. We can't give Museum of the Rockies full credit for Bozeman's economic success, but having a world-class museum in town doesn't hurt. Of course, it also doesn't hurt to have a major university, two ski resorts, a country club, six golf courses, and a billionaire.

In sharp contrast to Choteau, which makes do with a single grocery store.

*

As much as Ellie loves *Maiasaura,* she surely loves ice cream more. Thankfully, the Old Trail Ice Cream Parlor, on the edge of the museum, opens promptly at noon.

"Please . . ." she begs.

"One scoop," I concede, "once it opens."

Since we have half an hour to kill, Ellie settles into a picnic table alongside the museum's trio of nearly life-size dinosaur statues. Meanwhile, I open the van trunk to gather bread, peanut butter, chips, and drinks—all the fixings for our makeshift lunch. Back at the table, I notice I'd forgotten one item.

"Honey," I say, "you mind grabbing the jar of strawberry jam?"

"Sure," she says, hopping up to retrieve it from the cooler. No sooner does she grab it than it slips from her fingers, shattering glass across a six-foot radius of the museum parking lot.

I'd like to say I quickly adopted a "No sense crying over spilled jam" philosophy, proving myself at least as capable of caregiving as a mother *Maiasaura*.

Instead, I sighed loudly. A couple of times.

"How did it . . . just . . . slip?" I ask incredulously.

"I don't know!" Ellie says, her eyes filling like rain gauges. "One second, it was in my hands, and the next, it . . . wasn't."

I know, I know, accidents happen. But they seem to happen with far more frequency in my family. Collectively, we drop more dishware than a circus performer spinning plates. Given this daily practice of shattering, one might think I'd be better at making peace with all the pieces of busted glass. But the truth is, I'm a little perturbed. Maybe a lot perturbed.

"I'll be right back," I tell Ellie. "Just . . . don't touch anything."

The tinkling bell of the Old Trail Museum announces my return.

"Back again," Sean smiles.

"There's been an accident," I say miserably.

If you've never handpicked hundreds of shards of glass in a parking lot beneath Montana's beating sun, I do not recommend it. Thankfully, Sean dispatched me back to the lot with a roll of paper towels, which at least prevented me from cutting my hands.

As I gather the glass, a woman screeches toward us on her bicycle.

"Is that . . . jam?" she asks.

"Yeah," I say, "but don't worry, I'm cleaning it up."

"Well, be sure you do," she says, glancing nervously toward the wilderness. "It'll attract bears."

Suddenly I'm reminded of a few "Do Not Feed the Bears" signs I'd seen during the drive into town. Our humble pursuit of a peanut butter and jam sandwich is rapidly becoming a regional crisis.

Crouching low in the explosion's epicenter, I continue my long and tedious task, slipping shards of glass into the plastic bag. Then, I dab at the asphalt with paper towels. All the while, through her teary eyes, Ellie scans the trees for bears.

Parenting is hard, but so is being a kid. My failure is often in recognizing the latter point too late. At Ellie's age, making her parents proud is nearly as exciting as discovering a dinosaur fossil. And yet, too often, I'm so busy browbeating one kid or another for some perceived transgression that my unconditional love begins to feel a little too "conditional." It's a truth that leaves me feeling miserable. Plucking glass from a parking lot feels like the perfect penance.

A woman pops her head out of the soon-to-be-opened ice cream shop.

"Would a broom help?" she calls.

"A broom would be great," I admit.

She does me one better, handing me a broom and a bucket of soapy water. In bear country, there are procedures for these sorts of things.

Once the crisis is averted, Ellie and I make our way to the newly opened ice cream parlor.

"I'm sorry," she whispers.

"No, I'm sorry," I say. "That jam really got us into a jam, am I right?"

She snort-laughs.

"Honestly, I'm just glad we got out of that jam . . ."

"Stop!" she groans.

We sidle up to the ice cream shop counter, where the woman at the counter asks, "What'll it be?"

"A scoop of huckleberry ice cream," I say. And then, after a beat: "Better make it two."

*

Just up the road, the unincorporated community of Bynum is so small that we can hold our breath and drive from one side to the other. The main drag runs perhaps a third of a mile, though, along the way, we see all the major attractions: Susan's former school, the Trex Agate Shop, and the Montana Dinosaur Center—our fourth stop on the MDT.

The Dinosaur Center is easy to spot; it's the building with the giant *T. rex* displayed out front. Even if you somehow fail to see that, you can't miss the accompanying sign: "Home to the World's Longest Dinosaur."

We enter, anxious for our afternoon site tour but also to check out the advertised "world's longest dinosaur" skeleton model—a sauropod named Seismo, which measures 137 feet long. (Imagine four school buses parked end to end.)

Shortly after our neck-straining visit with Seismo, our tour guides gather Ellie and me alongside a family of four from New York. For the next two hours, the six of us will drive to undisclosed fossil-rich locations in the Montana countryside, where our guides will point out everything from bone fragments to egg shards.

"Now, before I get into my bone identification talk," one of the guides says as we exit the van, "I'm going to give you a little safety talk. We are very much in grizzly bear territory. Though I've never seen a bear on this program, it is a possibility. That's why I always point out where my bear spray is, right here on the right side of my backpack. Any questions?" the guide asks.

*Just one,* I think. *Is it a problem if I've got a jar's worth of strawberry jam scent embedded on my hands and clothes?*

*

Before leaving Bynum, we stop at the Trex Agate Shop—complete with its dinosaur-themed mural painted on the far side of the building's exterior. Upon entering, it becomes clear that the agate shop is a rock hound's paradise: chock-full of rocks, minerals, and turquoise stretched throughout the roadside store. Standing alongside the counter is sixty-eight-year-old Dave Trexler, son of Marion Brandvold, the latter of whom discovered the juvenile *Hadrosaur* fossils on Egg Mountain in 1978. Dressed in jeans and a button-up, Dave welcomes us heartily into his shop.

We introduce ourselves and say we'd love to hear his mother's story.

"Well, Mother was six years old when she found her first piece of dinosaur," Dave begins, adjusting his glasses. "That was 1917, the same year dinosaurs were first described from the Two Medicine Formation here in Montana." Six-year-old Marion was anxious to help her ranch family coat their cattle with creosote—a chemical meant to keep the flies away. But the ranch foreman, a cowboy named Bob Stonehouse, would have none of it. The work was too dangerous for a girl her age.

"Why don't you look for some pretty rocks on that hillside?" Stonehouse suggested.

When the work was done, Stonehouse wandered her way.

"Find any rocks?" he asked.

Marion pulled several from her pockets.

Stonehouse examined them, his eyes homing in on one. "Now this one here," he said, "used to be bone. And not just any bone, but a bone from an animal that no longer lives on the face of the earth. It's called a dinosaur."

Marion's eyes widened. She was hooked.

For decades, Marion roamed the Two Medicine Formation on which we now stand, looking for anything shiny, pointed, or fossilized. As a child, Dave regularly joined his mother on horseback rides through the sandstone terrain, their eyes firmly fixed on the ground.

At sixteen, Dave bought a 90CC Bridgestone motorcycle, which he hoped would help him cover more ground. Although Marion was initially opposed to the motorcycle, there was nothing better for fossil-finding transportation. Together, they spent long days riding across miles of ranchland. One afternoon, their search paid off. On a hillside, Dave spotted articulated dinosaur bones. He identified the fossils as part of a dorsal spine belonging to some dinosaur, but that's all he knew.

Since Marion and Dave wanted to learn more, in the summer of 1971, mother and son struck out on a road trip across Montana to visit every dinosaur displayed in the state.

Not that there was much to see.

"We saw one dinosaur skull and one partial dinosaur," Dave tells Ellie and me. The bulk of the other fossils had been shipped to museums out east.

"The further we went, the angrier we got," Dave explains. "We've got all these dinosaurs that Montana is famous for, yet we have exactly one on display."

Even the phrase "on display" seemed a stretch. The dinosaur in question was housed in the high school basement in the eastern Montana town of Ekalaka (population 600), which, at the time, Dave says, was little more than a town at the end of a gravel road.

But their visit proved inspiring. Marion and Dave determined that if Ekalaka could do it, Bynum could too. They set up a small museum, which today serves as the Trex Agate Shop.

Dave's goal was to find a complete specimen to showcase, and he and his mother continued their pursuit for years. In 1977 Dave discovered what he believed would be that dinosaur—a fossilized leg jutting from a hillside about half a mile from the nearest vehicle access point. Dave's wife Laurie discovered dinosaur skull fragments the following year, though much of the bone had been turned to powder. Their discoveries continued later that summer when Marion stumbled upon a few small vertebrae near the parking area. By day's end, they'd found a jaw to go with them.

While Dave quickly identified the jaw as belonging to a duckbilled dinosaur, its small size—coupled with incomplete growth patterns of various bones—led them to believe the fossils were from baby dinosaurs. Before this discovery, baby dinosaur fossils were virtually nonexistent. However, since Dave, Laurie, and Marion lived in rural Montana with no paleontology research institutions nearby, they had no way of recognizing the significance of what they had found.

It wasn't until Jack Horner—then a fossil preparator at Princeton University—and Bob Makela visited Trexler's rock shop later that summer that Dave, Marion, and Laurie realized just what rarities they had on their hands. Shortly after meeting Jack and Bob, Marion retrieved some of their newly discovered fossils and asked for their opinions on what they might be. Jack and Bob confirmed that the fossils belonged to a duckbilled dinosaur.

But, like Dave, the men were most surprised by the bones' size.

"The femur, or thighbone, of a typical duckbill might be four feet long and as thick as a fencepost," Jack Horner wrote. "The femur that Mrs. Brandvold handed me, if the bone had been whole, would have been the size of my thumb. . . . What I had in my hand was a bone of a baby dinosaur."

In the coming weeks, Jack and Bob obtained permission to dig at the site, where they soon discovered the remains of fifteen baby dinosaurs in a nest-like structure. As Jack and Bob excavated the nest site, Dave reported that his wife Laurie had found a large dinosaur skull two hundred yards from the nest. The Trexlers and Brandvolds had attempted to excavate the skull, but it had broken into several pieces, so they abandoned the site. After Jack and Bob finished excavating the nest, they unearthed the remains of the skull, discovering it to be a new species of duckbill they would later name *Maiasaura peeblesorum*. A study of the adult skull and the baby skulls revealed them to be of the same species.

"The fact that fifteen baby hadrosaurs had been feeding, and had stayed together for a period of time, indicates that some form of parental care was administered," Horner and Makela argued in their paper for *Nature*. They cited their belief that a parent likely brought food back to the nest, adding that if the babies had left the nest on their own, it seemed "unlikely that they would have returned to the nest together without parental supervision."

For many, it seemed a novel idea. In 1968 paleontologist Robert Bakker theorized that sauropods may have exhibited parental care by surrounding juveniles in a protective ring, though the idea was swiftly challenged. Yet by 1979, when Jack and Bob made their case for the *Maiasaura*'s parental duties, they had the proof to back it up—thanks, in part, to Marion Brandvold's discovery.

"The simple fact was those first duckbill babies were found, not by painstaking analysis on my part, but by sheer luck," Jack Horner wrote. "Marion Brandvold was lucky to find them. Bob and I were lucky to find her."

Working together, professional and amateur paleontologists changed our understanding of dinosaur behavior.

"Marion Brandvold had discovered a lovely little window on the Late Cretaceous," Horner wrote. "What we did was to open that window and climb through it."

*

The road between Bynum and our destination of Shelby is as straight as a proverbial arrow. Over breakfast, when I'd mentioned our route to Susan, she'd informed me that I could drive it with my eyes closed.

Not that I would. Not only for safety reasons but also because there's so much beauty to see along every mile of I-15 North. As I studied the map the previous day, it had all seemed so terribly far. Yet, while driving it, it doesn't seem far enough.

There's something inexplicably pleasurable about a long drive at dusk with your daughter. A stop-the-clock stillness takes hold. The plains stretch for seasons ahead of us, one after another in their endless, hypnotic array. Agricultural fields grip the landscape, both wheat and barley rooted to the ground. Somewhere beyond the window is the Marias River, the branch of the Missouri that the Lewis and Clark expedition smartly chose not to take.

The farther I drive, the more the landscape reveals its stories. The region's history—from westward exploration to the domestic lives of dinosaurs—is etched into the rock layers.

"Need anything?" I ask Ellie, who's deeply engaged in her audiobook.

"Huh?" she asks, slipping off her headphones.

"Do you need anything?"

"Oh. Nope."

"Pretty cool views out there, huh?"

She pauses to take it in, the way the prairies and grasslands seem to unfurl deep into the next century.

"Yup," she agrees. "That's a lot of land."

While chatting with Sean Doyle at the Old Trail Museum, he'd mentioned that Montana is roughly the size of Germany in terms of square mileage. "The difference," he'd added, "is that Germany has eighty million people, while Montana has just over a million."

Which explains the lack of traffic.

We arrive in Shelby at dinnertime, taking our seats at a picnic table at an outdoor diner. Ahead of us, dozens of railroad cars restock, refuel, and switch to the proper tracks. The town has a population of 3,100, and once, it was home to Jack Horner—who found his first dinosaur fossils not far from where we are now.

Over chicken tenders, we watch the train cars come and go. This is a place where things are perpetually moving, including a song sparrow, who hops atop our picnic table, angling for a french fry.

"Ellie," I whisper, "don't look now, but there's an avian dinosaur on the table."

Ellie turns, coming face-to-face with the curious bird.

"Hi, Dinosaur!" she laughs. "Want a fry?"

She tosses it one about the length of its body.

"Honey, no feeding the dinosaurs!" I chide. "Now, it'll never leave."

"But he's hungry! We gotta feed him!"

I think of the momma pig and her piglets. And the *Maiasaura*, Marion, Susan—and all the other parents and teachers who do all they can for their kids.

"Fine, fine," I sigh, turning toward the sparrow and tossing him one of my fries. "Help yourself, my friend."

FIG. 16. A fortuitous meeting with Jack Horner, one of America's most well-known paleontologists and the inspiration for *Jurassic Park*'s Dr. Alan Grant. Author photo.

# 8
# Hit the Road for Jack

### JULY 7

*Shelby,* MT → *Chinook,* MT → *Havre,* MT → *Rudyard,* MT → *Havre,* MT

For months, I'd tried to track down paleontologist Dr. Jack Horner. I'd sent emails and social media messages and even relied on a mutual friend. All my attempts were met with silence. Fearing that my outreach efforts might be teetering dangerously close to stalking, I did the reasonable thing and let the matter drop.

I certainly didn't blame him for his radio silence. As one of the most well-known paleontologists working today, I suspected the man was overwhelmed with interview requests. And the truth was, I wouldn't even know what to ask Jack if we had connected. Here was someone who'd threaded the needle between science and stardom, a man whose groundbreaking research had pushed beyond the pages of scientific journals to the big screen as a consultant for *Jurassic Park.* His findings have helped scientists and moviegoers alike see dinosaurs more clearly. And though he's yet to literally bring dinosaurs back from the dead, he resurrected them in our hearts and minds for a generation of folks like me.

The odds of meeting Jack Horner on our trip seemed slim. After all, our days were dwindling, and Horner appeared to be nowhere.

But those odds had dramatically improved the previous night, when—while lounging in our hotel room in Shelby—Jack Horner's birthplace—I received a phone message from Lila Redding, the curator of the Depot Museum in the nearby unincorporated community of Rudyard.

"B.J., we're hosting Jack Horner tomorrow night for a little wine and cheese fundraiser. I know you planned to visit us later in the week, but . . . well, Jack Horner will be here tomorrow."

If I could've hugged my phone, I would've.

"Who was that?" Ellie asked, her eyes never leaving the TV.

"Jack Horner's going to be in Rudyard tomorrow," I said, astonished by our good fortune.

"Who?" she asked.

"Jack Horner."

"Let me guess," she says. "He discovered some dinosaur or something."

"Yeah," I sigh. "Something like that."

*

In 1954 seven-year-old Jack Horner was roaming a ranch a few miles outside Shelby when he found his first dinosaur fossils. It was the beginning of a lifetime of successful fossil finds.

While school didn't always make sense to Jack, dinosaur fieldwork always did. His mother Miriam supported his love for fossils, regularly driving him and his brother to dig sites throughout Montana and beyond. Jack never had trouble reading the rocks; his struggles involved books and blackboards. Nevertheless, in 1964 Jack earned his high school diploma—with a D average.

The next fifteen years took Jack both near and far from home. First, he enrolled in Missoula's Montana State University (known today as the University of Montana) before being shipped on a fourteen-month stint in Vietnam as a Recon Marine. After returning to the states, he was in and out of the California Institute of Technology in short order, later returning to the University of Montana, where he flunked out a total of seven times.

While Jack struggled to graduate from college, that didn't prevent him from working at one. In 1975 Jack secured a position as a technician at Princeton University's Natural History Museum. While at Princeton, Jack uncovered the reason for his struggles with reading and writing. After reading a sign with questions that seemed written just for him ("Is reading difficult?" and "Would you rather watch a movie than read a book?"), Jack took the sign's advice and got tested for a learning disability. He was diagnosed with dyslexia.

While dyslexia may have hindered aspects of his formal education, today, Jack Horner credits his learning disability with helping him find his strengths. "From my failures, I've learned where I need help . . . But I've also learned from my accomplishments what I'm better at than the linear thinkers."

Though he struggled throughout his formal education, he has always excelled in the field. As did Jack's dig partner, Bob Makela—a high school science teacher in Rudyard with a degree in secondary education. Throughout the late 1970s and 1980s, Jack and Bob spent long days under the Montana sun unearthing fossils just outside the town. It didn't matter that neither man had a degree in paleontology, geology, or a related field. They learned from the land and each other.

Then tragedy struck.

On a June evening in 1987, while driving back from a dig, Bob Makela died in a car crash.

Despite the loss, Jack Horner regularly returns to Rudyard, where folks are always glad to see him.

As Ellie and I will see for ourselves later that night when we get to meet him too.

*

Since we've got a full day before Rudyard's wine and cheese fundraiser with Jack Horner, Ellie and I hit the road to collect two more Montana Dinosaur Trail passport stamps. Our first stop is the Blaine County Museum in Chinook (120 miles east of Shelby), followed by a twenty-mile backtrack to Havre, home of the H. Earl Clack Museum. Sure, we're logging a few more miles than our original itinerary, but it seems a small price to pay for a visit with Jack.

Driving into Chinook—the Milk River to the right—we spot a sign for Bear Paw Battlefield sixteen miles to the south. I'd long read of the famous final battle between the Nez Perce (who call themselves the Nimíipuu) and the U.S. Army, the latter of whom had cornered the tribe a mere forty-two miles from the Canadian border, where the Nez Perce sought refuge from

U.S. persecution. On October 5, 1877, after five days of intense fighting, Chief Joseph relayed his famous, albeit heartbreaking, message to the U.S. Army:

"Hear me, my chiefs. I am tired; my heart is sick and sad. From where the sun now stands I will fight no more forever."

The fighting stopped, and tribe members were forcibly relocated, first to Fort Leavenworth, then elsewhere.

Founded in 1880, Chinook was born in the aftermath of that battle. It initially served as the region's agricultural supply center. In 1887, with the addition of the Great Northern Railway line, the town evolved from a supply center to a railroad town. Chinook's population peaked at around 2,300 in the 1960s and has declined ever since.

Inside the Blaine County Museum, we're warmly greeted by museum director Samantha French.

Dressed in a black shirt and acid-wash jeans, the bespectacled twenty-eight-year-old appears at least a decade younger than the other museum directors we've met. She obtained her position in 2019, at the age of twenty-four, replacing her predecessor who'd held the role for a quarter century.

Though Samantha initially had limited museum experience, that's true of many small-town museum directors.

"You can figure it out along the way," Samantha assures.

Growing up in the nearby city of Havre provided Samantha an advantage due to her familiarity with the area. However, even though Havre was home, it never truly felt that way while she lived there.

"Not to say that small towns are unfriendly, but if you never quite fit in, you never quite fit in," she explains. Throughout high school, Samantha was more interested in art and history than Friday night football, though she finally found her fit while attending college in Helena. She majored in history, interned at the state historical society, and, through these experiences, acquired the skills that serve her in her current role.

Yet before settling in Montana, she had to leave it—embarking upon a one-year master's program in art history at the University of Bristol in southwest England. Upon returning stateside, she viewed northern Montana with fresh eyes.

"Growing up here, I never liked it," she admits. "There's a reason I wanted to go to the UK. I used to think that everything good in the world was elsewhere, that all the best museums were in big cities. I was always looking toward other places."

However, that changed when Samantha accepted the museum directorship in Chinook.

"I've learned our area is really special and really important. And I want to make sure that people who are like the old me can understand that there's a history here that's valuable."

I recall what Sean Doyle shared in Choteau: that, considering the harsh winters and the vast distances between towns, you really have to want to live in Montana. Although Samantha didn't—not at first—now she's one of central Montana's greatest defenders.

Samantha gives us a tour, beginning with the museum's signature dinosaur, a cast of the *Gorgosaurus* skull behind glass. This cast was taken from a *Gorgosaurus* found in Alberta, Canada, though it represents teeth and vertebrae found within miles of where we are now.

"Did you hear about the *Gorgosaurus* found north of Havre?" she asks. "It was sold at Sotheby's to a private collector. But it goes to show there's a lot in this area that has yet to be found. And who knows what the next significant finding will be? If somebody were to give such a find to our museum," she continues, "that would help put us on the map."

*

Back in Havre (population 9,362), Ellie and I enter the H. Earl Clack Memorial Museum—our sixth stop on the MDT—where Lela Patera, a museum volunteer, greets us.

"So glad you're here!" Lela calls. "And I don't know if you've heard, but Jack Horner will be in Rudyard tonight!"

FIG. 17. Ellie poses with Samantha French, director of the Blaine County Museum in Chinook, Montana. Author photo.

(Jack's visit to Rudyard isn't just the talk of the town; it's the talk of the entire state.)

"We just got the message," I say. "We can't wait."

"Jack's great," Lela says. "We were at Montana State at the same time. I even went to a party at his house once," she continues, leaning in close, "*before* he was famous."

Jack Horner's hardly the only celebrity in Lela's orbit. Just ahead of us, encased behind glass, is the 74-million-year-old skull of Lamby the *Lambeosaurus*—the museum's signature dinosaur. A member of the Hadrosauridae family, this duckbilled herbivore is distinguished by its hollow cranial crest, complete with nasal passage. While the crest's function remains unclear, some theorize it served a social function, trumpeting its desire for courtship. Others suggest it enhanced the animal's sense of smell. Whatever its role, the unmistakable bony protrusion is an easy identifier for Ellie and me. If it's beaked with a crest, odds are it's a *Lambeosaurus*.

As Lela explains, Havre's story extends beyond its dinosaurs. Like many northern Montana towns, it was founded to support the Great Northern Railroad. When railroad director James J. Hill first conceived of his transcontinental line, he hoped to build it from the bankrupt lines that came before. After purchasing the defunct Saint Paul and Pacific Railroad, he and his partners expanded the routes.

"What we want," Hill remarked, "is the best possible line, shortest distance, lowest grades, and least curvature we can build. We do not care enough for Rocky Mountains scenery to spend a large sum of money developing it." Thus, Hill's western route steered north of the mountains, making it the northernmost transcontinental line in the country. Given the rail line's positioning, the 250 or so miles of track between Havre and Whitefish was given the nickname the "Hi-Line."

While the Great Northern Railway greatly contributed to the development of Havre, it also created several other small towns along the way.

"Most of the towns west of here are spaced seven miles apart," Lela explains. "And that's the result of the railroad. Those steam engines needed water about every seven miles."

Later, when I check the map, I'll see Lela's right. Heading west from Havre—from Burnham, to Kremlin, to Gildford, to Hingham, to Rudyard, and beyond—all these towns are, in fact, separated by seven or so miles.

When you stop to learn its stories, the landscape has a logic. Everything—from where a dinosaur dies to where a railroad thrives—is guided by the land.

And in the case of railroads and roads, nothing beats a straight, flat line.

*

Having never attended a wine and cheese fundraiser in small-town Montana, I'm left guessing at the proper attire. While Ellie can pull off any look, I dedicate part of the afternoon to browsing through Havre's secondhand stores in search of a collared shirt.

"Ugh, nothing here works!" I lament with the melodrama of a prom-bound teenager. "Oh, wait," I say, pressing a mud-colored flannel to my chest, "how about this?"

"Ummm . . . it's brown," Ellie says.

"In a good way?" I ask.

"In a brown way," she says.

At last, I unearth a mostly innocuous navy-blue collared shirt that doesn't look like I'm swimming in it.

"Not my style," I tell Ellie as I make my way toward the register, "but it'll have to do."

As we walk toward Rudyard's Depot Museum later that evening, a cursory skim of the crowd confirms that even a collared shirt is more than I require. In Rudyard—much to my delight—blue jeans are the new black tie.

While the Depot Museum is home to many treasures, it has earned its place on the Montana Dinosaur Trail thanks to its fully articulated *Gryposaurus*. Similar to the *Lambeosaurus* in Havre, the *Gryposaurus* is also a Late Cretaceous member of the Hadrosauridae family. Named for its hooked nose, this herbivore roamed the region seventy-five million years ago. The people of Rudyard have nicknamed their *Gryposaurus* "the oldest sorehead." However, it's not the only "sorehead" in town.

Just outside of town, a billboard proudly proclaims Rudyard home to "596 nice people and 1 old sorehead." The "old sorehead" refers not to the dinosaur but to the town's "grumpiest" elder. Following a town vote (which involves placing a couple of crumpled bills into a coffee can featuring the candidate's name and photo, with all proceeds going to the local senior center), the "winner" begrudgingly accepts the "honor" of being "the old sorehead." Yet when I reference this unofficial town motto to a woman working the museum cash register, she says, "Yeah, well, we wish we still had 596 people living here. These days, it's about half that."

Despite Rudyard's dwindling population, every citizen seems to be in attendance tonight, roaming the dinosaur portion of the museum, as well as the various other structures on the museum grounds. For a town of 270, the museum punches well above its weight. Indeed, it's the best collection of local history we've seen anywhere. Directly behind the dinosaur building is a nineteenth-century home chock-full of local paraphernalia (from banners to baseballs to military uniforms), while just beyond, a barn-like structure displays what I can only assume is one of the world's largest barbed wire collections, in addition to nearly three hundred lanterns, tractors, toys, sleds, and a Civil War saddle. Nearby, a windmill spins in the museum's courtyard, where a woman desperately tries to corral local children into a group photo.

"Come on, kids, let's get in two rows," she wrangles. "Two rows, please."

This isn't just a group photo. Unbeknownst to us, we've stumbled into a Klondike Bar publicity campaign that asked towns across America to answer the question, "What would your hometown do for a Klondike Bar?"

Drawing on Rudyard's "old sorehead" history, the woman with the camera has choreographed an answer to that question: a "young sorehead" (i.e., a small child) will hand a Klondike Bar to the "old sorehead" (i.e., a guy named Bob Christianson). God willing, this "passing of the Klondike Bar" will be captured on video, along with a photo of smiling children munching their ice cream treats.

Spotting Ellie and taking a "more the merrier" approach to the town's homegrown campaign, the woman asks, "Do you want in the photo, sweetheart?"

"Um . . ."

"Here, let's get you a Klondike Bar . . ."

Amid all her corralling, the woman hands Ellie an ice cream bar. Then, when the children are sufficiently posed, a pair of townspeople hidden inside two giant inflatable *T. rex* costumes bop into the frame alongside old sorehead Bob.

Were it not for my "old sorehead" research, I might've lost the thread of what is happening here. (I suspect that the good people at Klondike may not have fully understood the "sorehead"-related nuances either.) But as publicity stunts go, Rudyard has a pretty good track record. In 2017—just months after NBC's *TODAY* featured Susan Luinstra and her dancing pupils in Bynum—the morning show also featured that year's "old sorehead" election in Rudyard. Suffice it to say, it was a good year for small-town Montanans.

And tonight's a good night for us. After all our Montana miles, fate has rewarded us by allowing us to witness the warmth of this town at its best. Better still is that Ellie—having spent all of ten minutes here—is now a full-fledged Rudyardian, endowed with all the rights and privileges afforded its citizens (read: free ice cream). Were it not for the MDT, I'd have never heard of Rudyard, let alone visited. I'm so grateful that we did.

Following the photo, Ellie and I converse with a few locals.

"What's it like here?" I ask.

"Well, like a lot of small towns, we're dying off," the first woman shares. "We've been slowly dying off since World War II."

"When we first came here sixty years ago," her friend says, "there were probably three to four times as many people."

In the time between, population decline diminished the once-thriving community. By now, it's become a familiar story. There was a time when this unincorporated community was home to a couple of grocery stores, an auto parts store, tire stores, a dance hall, a movie theater, a café, and the best hardware store in the area. While the movie theater is still in business (on our drive into town, I'd noticed the paint-peeling marquee noting its double feature: *Jurassic World* and *The Land Before Time*), it's closed more nights than it's open. Today, it's a throwback to yesteryear, when families filled the seats after long days of working the land.

Over time, a generation of farmers left town or died, and few folks came to replace them.

Yet, as we observe tonight—as a fiddle band plays just behind us—the townspeople who remain take pride in the place they call home.

No one more so than Lila Redding, the unofficial mayor of Rudyard, who sits on a bench inside a barn-like structure overflowing with lanterns, tractors, and barbed wire. The exhibitions are pristine and greatly admired by the locals, whose boots click along the cement floor. This section of the museum complex proves that Rudyard is more than dinosaurs; it's a place that knows and values its history—even as it struggles not to become history itself.

"I can't thank you enough for calling us about Jack's visit," I say, sitting beside her on the bench.

Lila smiles. "A few days ago, Jack turned to me and said, 'Do you realize we've been friends for nearly fifty years?'"

"That's a long time."

"We go way back," Lila says. "Jack's been digging on our property for years. He just shows up," she chuckles. "Jack has the key to my house."

While some people have forgotten Rudyard, Jack hasn't. For decades, he's led dig trips on the dinosaur-rich Redding property twenty miles from the unincorporated community.

Today, Rudyard's centerpiece might be where we are now—the Depot Museum, an idea born in 1994 over cups of coffee at the local café.

Lila's father and friends were engaged in their usual banter when their waitress remarked, "You know, the Depot's for sale for a buck. You all should buy it and start a museum."

The men chuckled. One of them even handed her a dollar and wished her luck.

Not only did the waitress purchase the property with their dollar, but she also helped raise the funds to move the former train depot to its current location. The "wild idea" suddenly became a reality, and the five men quickly joined the effort, donating their lifetimes' worth of accumulated town-related artifacts.

Next, they moved an old country schoolhouse onto the museum property. After discovering the *Gryposaurus*, Jack Horner advised Lila and others to consider building a museum befitting their town's greatest paleontological find.

"I sent letters to people from around here who've done extremely well with their lives," Lila says. "People with money." And many people responded to the call by opening their checkbooks.

Lila raised $40,000—enough to purchase the necessary supplies.

"Then," Lila says, "we had a barn-raising for the next two weekends."

Or a museum-raising, more accurately.

Over four days, the townspeople built the museum from scratch.

"This museum has helped keep our economy going," Lila says. "We used to do two hundred visitors a summer, and now we do over a thousand."

In June alone, the museum boasted over four hundred visitors, nearly double the population of Rudyard.

Being on the Montana Dinosaur Trail helps, Lila agrees.

But so does hosting Jack Horner.

*

Inside the museum, we find Jack Horner seated at a table near the *Gryposaurus*, just feet away from a tribute to his late research partner Bob Makela. Jack listens politely as a man gesticulates what appears to be a rather dramatic story. Jack nods and responds, then holds out his hand and thanks the man for chatting.

Summoning my courage, we make our move, though not before a bit of cowardice slips in.

"After you," I say, nudging my nine-year-old forward to break the ice.

I expect a big-shot movie star type; I find a kind and humble man instead. In retrospect, maybe that's one of the qualities I admire most about him. Not "just" his paleontological discoveries (too many to enumerate here), but that a person can stay grounded even when their achievements have vaulted them into the scientific stratosphere. In the months leading up to our trip, I'd reached out to various scientists in the dinosaur world, some of whom responded to me with great enthusiasm and others who allowed my emails to wither in their inboxes. While Jack and I had missed connections a time or two, the error was my own: my outreach attempts went to incorrect email addresses, and my messages wound up in social media accounts he didn't check. The best way to meet Jack Horner, I've learned, is to meet him face-to-face.

Preferably in a place like Rudyard.

"Hi there," seventy-seven-year-old Jack smiles. He's bearded, wearing a blue button-up shirt and a red baseball cap.

"Hi," Ellie bashfully replies.

"What brings you two to Rudyard?" Jack asks.

"We're on the Montana Dinosaur Trail," I say.

"Oh yeah?" Jack says, his interest piqued. "How are you liking it?" he asks Ellie.

"Good," Ellie says. "I love it."

"I bet you and your dad are having quite an adventure," he says.

"Uh-huh," Ellie agrees.

What I want to say is: *Excuse me, Jack, but can I borrow you for an intensive three-hour interview?* Instead, I whisper to Ellie, "Do you want to see if he'll sign your book?"

I hand her our newly purchased copy of Jack's latest contribution, a children's book titled *Lily and Maia: A Dinosaur Adventure.*

"Will you sign this?" Ellie asks.

"Sure," Jack agrees, uncapping his pen. "Who's it for?"

"Eleanor," Ellie says, preferring her formal name for such occasions. As Jack gets to work on the inscription, I lean toward her. "Do you want to see if he can make it out to your brother and sister too?" I whisper.

"No," she whispers in reply.

I try another tack.

"But wouldn't it be nice if you did?"

"Fine," she huffs, turning toward him. "And could you please sign it for my brother and sister too?"

Jack's happy to oblige, carefully repeating the spelling of each name aloud before committing it to ink.

"There you go," Jack says, returning the book to her. "And remember," he says, leaning in close, "this book is just for you, okay? Tell me what you think of it."

Ellie promises.

"How old are you anyway?" Jack asks.

"I'm nine."

"You know," Jack muses, "that's about the age Lily is in the book. She goes on an adventure by herself. Can you imagine going on an adventure by yourself?"

"We can barely manage it together," I joke.

Smiling, Jack turns his attention toward me. Mindful not to monopolize his time, I ask a single question.

"Is there one thing," I begin, "that most folks don't understand about dinosaurs?"

"Well," he says, capping his pen, "most people don't realize how long an interval of time there is between dinosaurs and humans. And they don't understand that mammoth elephants aren't dinosaurs. Even the simple stuff—the basics—some people don't know."

I shake my head as if to imply that I am not one of those people.

"A lot of people don't think the history of the world is important until things start happening," he says. "That's when they realize that the past really is the future."

He doesn't specify just what "things start happening" means—he doesn't have to. We are currently experiencing the hottest summer on record.

"It's important," Jack continues, "to understand the evolution of ecosystems through time and react to them."

Outside, a stiff wind blows the windmill blades as dark clouds begin clotting the sky.

"We drove a long way to be here," I say, wrapping an arm around Ellie, "and I'm so glad we had the chance to meet you."

"And I'm so glad you came," Jack says.

*

After our meeting with Jack (and after securing our seventh passport stamp), we hightail it back to Havre. The sky had turned dangerously dark, and a storm threatened throughout our nighttime drive. Yet somehow, we were spared the downpour. Though the van could've used a good rinse, its red paint disappeared behind a thick layer of Montana dust. If we wanted to get the van clean again, we'd have to take matters into our own hands.

Thankfully, just past our motel, I spot a sign for a do-it-yourself car wash.

"You up for one more stop?" I ask.

"Is it a museum?" Ellie asks worriedly.

I park atop the drain grate in the wash bay and feed five bucks' worth of quarters into the machine. I listen as the thrum of the pressurized water gun hums to life, then watch as Ellie's eyes widen with glee.

How to describe what happens next? How I spray that dust to smithereens, then scrub a layer of soap along every square inch of the van. And how Ellie cheers me on from the safety of her seat, her fists pumping the air while I dance to a song that no one else hears. Ellie, still buckled, shows off a few of her moves—smushing her face to the glass and inflating her cheeks half the size of balloons. We are as goofy as we are sleep-deprived, happy and homesick too.

Nearby, a digital clock ticks down the seconds until our time is up.

When it's done—once the van is back to looking its best—I buckle up and turn toward Ellie.

"Well?" I grin. "It sure beats a museum, huh?"

"Oh, yeah!" she says.

If only there were a passport for car washes.

FIG. 18. A *Triceratops* skull fossil like the one discovered by Charles Sternberg and sons in Wyoming in 1908. Sternberg, George Fryer 1883–1969, "074-04: Triceratops Skull Fossil" (2021). *George Sternberg Album #1 - Early Pictures*. 325, Forsyth Library Special Collections, Fort Hays State University.

# 9
# Leonardo and *Matilda*

## JULY 8

*Havre,* MT → *Malta,* MT → *Fort Peck,* MT

From July through September 1908, Charles Sternberg—the former longtime assistant to Edward Drinker Cope—scoured the sandstone and clay of eastern Wyoming in search of a *Triceratops* skull. He was joined by his three sons (George, Levi, and Charles Jr.), but the Sternbergs had little luck.

"Day after day hoping against hope we struggled bravely on," Charles Sternberg wrote. "Every night the boys gave answer to my anxious inquiry, What have you found? Nothing."

Following Edward Drinker Cope's death in 1897, Sternberg continued his search for dinosaurs. Only now, rather than serving as an assistant, he was the leader of his party, which came with both perks and perils—especially when the men he led were his sons.

The Sternbergs were tired, hungry, and sixty-five miles from camp. Their fossil-hunting season was nearing its end. Though Charles Sternberg had struck a deal to sell any newly discovered *Triceratops* to the British Museum of Natural History, it seemed 1908 would yield little in the way of a paycheck.

Until one August day, while roaming about the reddish shale near an ancient peat bog, Sternberg struck upon the horn core of the creature in question.

"How thankful we were that after so much useless labor," Sternberg wrote, "we had at last secured the great object of our hunt." Though it was hardly a perfect specimen, it was deemed a success given that *Triceratops* skulls were extremely rare. And it couldn't have come at a better time. As Charles Sr. and Jr. worked to unearth the skull, George and Levi, in a nearby camp, were subsisting solely on boiled potatoes. It was time to return to base. Or so Sternberg thought. But upon reuniting, George and Levi shared another find with their father and brother—this one even more impressive than the *Triceratops*.

"Shall I ever experience such joy," Sternberg wrote, "as when I stood in the quarry for the first time and beheld lying in state the most complete skeleton of an extinct animal I have ever seen?"

Initially identified as a *Trachodon* (today, it's known as an *Edmontosaurus*), Sternberg hailed the find as "the crowning specimen" of his life's work, not only for its completeness but for its mummification. Unlike the usual fare of dinosaur fossils, dinosaur mummies are unique in retaining well-preserved traces of skin and, on occasion, tendons and internal organs. Their research potential is unparalleled.

"It lay there with expanded ribs as in life," Sternberg wrote, "wrapped in the impressions of the skin whose beautiful patterns of octagonal plates marked the fine sandstone above the bones."

Sternberg's dinosaur mummy has been displayed at the American Natural History Museum for more than a century. To this day, it remains one of only a handful of dinosaur mummies ever found.

Another resides at the Great Plains Dinosaur Museum in Malta, our eighth stop on the Montana Dinosaur Trail. Known as Leonardo, the duckbilled *Brachylophosaurus* dinosaur mummy is so popular (and can command such a hefty loan fee) that the museum has loaned its star attraction to Japan. On this day, what Ellie and I are looking at is a replica of Leonardo. Though for a guy like me who can't tell the difference, I only know what I'm seeing looks pretty weird. Not a dinosaur fossil, but something closer to a dinosaur corpse. Leonardo's skin sheaths much of its body, its legs bent in what appears to be a mid-gallop. Indeed, there is something almost equine in its pose, something wonderfully reminiscent of the modern world.

After securing our passport stamp, Ellie and I walk across the Great Plains Museum parking lot to the Phillips County Museum, where we'll see our second *Brachylophosaurus* of the morning. For a town of 1,800 to possess a pair of *Brachylophosauruses* within spitting distance seems a stroke of good luck. And yet, for Ellie, it feels like bad luck.

"Not another one," she whispers as we enter the second museum moments after leaving the first.

Under different circumstances, such paleontological privileges would serve as the high of our day. But we are on day ten of nonstop dinosaurs, and though I'm ashamed to admit it, Ellie and I have grown a bit weary of the ongoing parade of ancient marvels. Give us a living, breathing dinosaur, and we'll take note, but we've begun to have our fill as far as fossils go. It's a stark reminder that our love for dinosaurs has its limits—which is something you'll never hear from a paleontologist.

We can't blame the dinosaurs or their museums. Our road-weariness is the inevitable result of jam-packed days filled with too many miles and not enough bathroom breaks. It's the result of too much uninterrupted father-daughter time and not nearly enough time for ourselves. After ten days in minivans, tents, and the occasional motel room, our close quarters have grown uncomfortably close.

This is not to say that we aren't enjoying ourselves. But I get the sense that Ellie wouldn't mind leaving the dinosaurs in our dust for a while.

At the Phillips County Museum, I peer down at Elvis, the thirty-three-foot-long *Brachylophosaurus* found on a hillside fifteen miles north of Malta. While Elvis is no mummy, it is 95 percent complete—hardly chopped liver.

I admire Elvis's bones for what seems like an appropriate amount of time. I yawn. I check my watch.

"Seen enough?" I ask Ellie a few minutes in.

But Ellie's no longer beside me. Instead, she's already back at the gift shop, acquiring her ninth passport stamp.

She waves me back toward the van.

"Come on," she says. "This passport won't fill itself."

*

While Ellie and I don't appreciate Leonardo the dinosaur mummy as much as we should, to modern researchers, it is the gift that keeps on giving: a near-perfectly preserved window into the Late Cretaceous. Leonardo's skin, soft tissue, and stomach contents allow scientists a wealth of information, including what Leonardo had for his last meal—a delightful pairing of ferns, conifers, and magnolia-like plants.

Dave Trexler (from whom Ellie and I first learned of his mother's discovery of baby *Maiasaura* fossils during our visit to the Trex Agate Shop back in Bynum) oversaw the scientific team that studied Leonardo from 2002 to 2008. He'd been called to help an amateur paleontologist who'd been running Malta's Dinosaur Field Station. Since Dave was one of the only professional paleontologists working with the Montana Dinosaur Trail's smaller museums, calling him made sense. And indeed, Dave was glad to have picked up the phone.

"Leonardo was just so spectacular," he shared with me. "It was the first large dinosaur found with a fully encased mummified body." While Dave estimates that a dozen or so dinosaur mummies have been found, most of those had been ruptured, meaning their internal organs were no longer in place.

"What Leonardo offered," Dave continued, "was the first time we could show that whatever was remaining inside that fossilized animal body was what was in the body when the animal died."

In February 2008, after years of preparation, Dave was called to Malta's newly opened Great Plains Dinosaur Museum to offer an independent assessment on whether Leonardo could be safely transported to NASA's Johnson Space Center in Houston, home to a computed tomography (CT) scan large enough for a *Brachylophosaurus*. What better noninvasive way to see inside than with a scan?

But it took Dave no more than a glance to determine that Leonardo had been "overprepared" by its many preparators. Much to Dave's frustration, Leonardo's plaster jacket had been cut away from the edges of the block, making safe transport all but impossible.

This proved a problem given that tens of thousands of dollars had already been invested in transporting the dinosaur mummy to Houston for scans.

Leonardo was scheduled to move in three days, but in its current state, Dave worried that it would never survive the journey.

"We had Hydrocal plaster that would work to rejacket it," Dave said, "but the plaster was over at my facility 251 miles away," in Bynum.

One of Dave's staff members made the late-night drive to deliver the plaster, and for the next forty-eight hours, Dave and others worked round the clock to rejacket and stabilize the fossil, which had been insured at over $2 million.

After remaining perfectly preserved for over seventy-seven million years, the prospect of destroying Leonardo en route to his close-up seemed an indignity too great to bear. It was transported by semi-truck on an undisclosed route, Dave Trexler sitting shotgun all the way. Fortunately, Leonardo survived its cross-country adventure thanks to its rejacketing and careful packaging.

Dave joined other paleontologists for several days to further prep Leonardo for its scans. However, upon arriving at NASA, the team learned that the CT scan machine was currently inoperable; thus, they made do with an x-ray scan.

The results yielded an interior look at a once-living dinosaur—from what it ate to how it ate it. However, a few mysteries remain without a CT scan—which provides 3D imaging.

"If we can get the right equipment and personnel, we could learn more about what these animals actually were like internally," Dave explains.

Another way to learn more about Leonardo is to perform an invasive autopsy—digging through the soft tissue to see what's within. However, given the advancements in paleontological technology, Dave strongly opposes such action.

"It's always been my preference that we don't destroy this one-of-a-kind specimen," Dave explained. "We owe it to the specimen and to future generations to wait and do this non-destructively. Once you take something apart," Dave says, "you can never put it back together again."

*

Following our visits to both Malta museums, Ellie and I buckle up en route to Fort Peck, our last stop of the day.

Fort Peck began as a trading post in 1867, though it transformed into a town in 1934. It was built for the U.S. Army Corps of Engineers, which, at the behest of President Franklin D. Roosevelt, was tasked with constructing the Fort Peck Dam as part of the New Deal. At a price tag of $100 million (approximately $1.8 billion in 2025 purchasing power), the dam was one of the era's most expensive (and technical) projects. Intended to improve navigation and prevent flooding, a not-so-subtle additional objective was to employ more than fifty thousand American workers in desperate need of jobs.

Today, Fort Peck is home to 239 residents, many of whom work at the dam or the accompanying power plant. Yet, upon driving into town, we're greeted by one of the U.S. Army Corps of Engineers' lesser-known projects: the Fort Peck Theatre. Constructed in nine months in 1934, the Swiss chalet-inspired theater served as a much-needed source of entertainment throughout the dam's construction. The theater premiered its first film, *The Richest Girl in the World*, in November of 1934, and for the next several years, according to the theater's website, played movies "24 hours a day, seven days a week." Dam workers and their families waited as long as necessary to take their places among the theater's 1,209 seats. There, beneath the seventy-four-foot-long hand-hewn wooden beams, they'd transport themselves to far-off lands through the magic of the movies.

For several decades, the Fort Peck Theatre has been home to the Fort Peck Summer Theatre, which produces several theatrical productions each summer.

How a town of 239 runs a summer's worth of shows is beyond me. I'd hoped to see one myself, and upon browsing the website in the weeks leading up to our trip, I was thrilled to learn that Roald Dahl's *Matilda*—one of Ellie's favorite books—would be performed on the night of our visit.

I'd snagged a pair of front-row tickets and kept the show secret for weeks.

Though as we drive past the theater late that afternoon, a giant roadside marquee announces the evening's show.

I sweat as Ellie comments, "What a cool theater!"

"It sure is," I say, speeding up, hoping she won't notice the marquee.

Minutes later, when we check into the historic Fort Peck Hotel—another 1930s-era structure built in the Swiss chalet style—the woman behind the front desk innocently asks, "Are you two going to *Matilda* tonight?"

"Okay then," I say, reaching for the room key and ignoring her question wholesale. "Thank you so much!"

"Wait . . . what did she say?" Ellie whispers as we lug our luggage up the hotel's wide stairway.

"Nothing," I say, "not a thing."

I can only credit Ellie's momentarily lapses in reading and listening skills for keeping my secret alive. In this instance, aloofness serves her well.

We enter our room, which—much like the hotel itself—is a time capsule from the 1930s. Quilt-covered twin beds occupy space along adjacent walls opposite a sink. The walls are the color of sky, the carpet the color of Garfield. A bureau and mirror set appear pulled straight from an antique store. I suspect it's been there for at least half a century. Exiting the room, we spot a long hallway on either side, its walls filled with play posters from previous seasons.

"I wonder if there are any plays tonight," Ellie asks.

"I doubt it," I lie.

"Not on a Saturday night?" she asks.

I turn toward her, an eyebrow raised.

"Since when do you know what day of the week it is?" I ask.

Later that afternoon, after a swim in a nearby lake (which we bill as a search for *Plesiosaurs*), we return to the room and scour our suitcases for fresh clothes.

"Why don't you wear your nice shirt tonight?" I ask.

"Why?" Ellie asks, her hair still flecked with water.

"I don't know," I shrug. "Sometimes it's just nice to dress up."

Ellie senses trouble.

"We're not going to another wine and cheese thing, are we?"

I assure her we are not.

"Or"—she gulps, terror in her eyes—"another museum?"

"Honey . . . no. No museum."

"Well," she says, deducing our destination like a young Miss Marple, "then I guess we must be going to a play."

One look tells me she knows everything.

"How long have you known?" I ask incredulously.

"About the play?" she asks, reaching for her halter top. "I don't know. A little while. There was that sign when we drove past the theater. Plus, that lady mentioned it at check-in."

"And you were just . . . playing dumb this whole time?" I ask.

"I mean . . ." she says, "I knew you wanted it to be a surprise."

I make my way across the room to wrap her in a hug.

"Oh, honey," I say, "that you cared so much about *my* surprise makes for an even better one."

Over the past ten days, we've driven through a good chunk of our country. We've seen countless fossils and even dug up a few ourselves. We've waded in the Yellowstone River. Learned about motherhood from some pigs and a *Maiasaura*. Sheepishly yawned over a mummy. But of everything we've done—and everything we'll do—nothing brings me closer to my daughter than my surprise overshadowed by her own.

Ellie and I stroll from the hotel to the theater a few minutes before showtime. We are joined by the entire town of Fort Peck and much of the surrounding area. The parking lot is packed with pickup trucks, their plates revealing names that read like fairytales.

"Saskatchewan?" Ellie says, studying the plate closest to it. "Where's that?"

"Somewhere in Canada," I say vaguely.

(Seven hours to the north, I'll later learn.)

It seems a rather long commute for community theater, but then again, given the miles we've logged for fossils, who are we to judge?

Upon taking our seats in the far-left front row, all those miles suddenly seem a small price to pay.

As the lights dim, the crowd crackles with excitement. And when young Matilda leaps to the stage and breaks into song, I can barely keep Ellie in

FIG. 19. Ellie outside the Fort Peck Theatre before a "surprise" viewing of *Matilda*. Author photo.

her seat. There is so much life on that stage that it spills into the audience. It washes over us like a wave of electric current, sparking, dazzling, and daring us to blink.

At intermission, we forgo the long concession stand line and return to the van parked in the hotel lot. I open the trunk, and we take our seats on the tailgate, helping ourselves to prepacked snacks and bottled water—provisions I'd *Tetris*-ed into the trunk nearly two weeks prior.

Ahead of us, we can just make out the edge of the theater, now lit with marquee lights.

"How do you like it?" I ask.

"I love it," she says between swigs.

"Who's your favorite character?"

"Miss Trunchbull," she says. "No, wait, Matilda's dad."

"They're pretty funny, huh?"

"*Super* funny," she laughs. "The funniest." She returns the cap to the gallon of water and turns toward me.

"Dad?" she says.

"Yeah, hon?"

"Thank you for bringing me here."

Together, we peer out at the fading Montana light as the big sky drops its dark curtain.

"You're welcome," I say. "Honestly, hon, I've loved every moment of this trip."

"No," she says, rolling her eyes. "I mean, thank you for taking me *here*. To the theater. To see *Matilda*."

"Oh, right," I say, searching for my stiff upper lip. "No, I know. Plays are great."

"Come on," she says, taking me by the hand, "or we'll miss the second act."

As we head toward the theater, she hums our new favorite song: "Take Me Home, Country Roads."

Like it or not, our show is more than half over. But it's not quite over yet.

FIG. 20. Ellie and Natural Resource Specialist Sue Dalbey pose alongside a *T. rex* at the Fort Peck Interpretive Center. Author photo.

# 10
# Amateur Hour

### JULY 9

*Fort Peck, MT → Jordan, MT → Glendive, MT*

On September 5, 1988—Labor Day—thirty-five-year-old Kathy Wankel, her husband Tom, and their two young children were enjoying the three-day weekend with a camping trip at Fort Peck Reservoir. Kathy had always been a bit of a rock hound, and back at the family ranch, she could often be found scanning the land in search of fossils, minerals, or bones. This is what she was doing that morning, around 9 a.m., when the light fell upon a bone protruding from the ground.

That late into the summer, drought season had significantly lowered the lake, revealing more shoreline and, with it—Could it be? Was it possible?—what appeared to be a shoulder blade belonging to a *Tyrannosaurus rex*.

The following month, the Wankels returned to the site with more time and tools. The family carefully removed a few fossils from the hardened clay, transporting them back to the ranch in a cooler. Kathy rinsed them in the back porch sink before displaying them on the Ping-Pong table, where they remained for the next month.

In late November, Kathy and Tom drove the bones to Bozeman's Museum of the Rockies for an expert opinion on what they'd found at the reservoir. They parked in the museum's loading dock, where the chief fossil preparator met them.

The preparator looked at the bones inside the cardboard trays in the back of the station wagon, and then said, "You'd better come inside."

*

I wake shortly after sunup, reaching for a stack of postcards I've been meaning to write all week. In the bed at the far end of the room, Ellie sleeps as only a nine-year-old can, so deep that I place my hand by her mouth to confirm she's breathing. Proof of life secured, I creep down the nearly hundred-year-old hotel stairs before settling into the screened porch that runs the length of the back of the hotel.

The previous night, following *Matilda,* Ellie and I had all but taken up residence there, staring out into the moon-drenched dark as a few hotel guests sidled up to the interior bar.

Seated in rocking chairs, I'd read aloud from Louis Sachar's *There's a Boy in the Girls' Bathroom*—a book starring fifth grader Bradley Chalkers, a boy so utterly annoying ("Give me a dollar or I'll spit on you"), that he understandably has a hard time making friends. That is, until a new boy comes to town who doesn't know enough about Bradley's reputation to write him off completely.

Over the past two days, Ellie has asked me to read from the book during every spare second we've had. While she's enjoyed *Ramona the Brave* and *Bridge to Terabithia,* this book hits her smack dab in the heart.

She listened intently for half an hour that night; her eyes closed as we arrived at the scene where Bradley drew a monster while chatting with the school counselor.

When Bradley asks the counselor if people can see the "good" inside monsters, the counselor confirms they can.

"'Well, how does a monster stop being a monster?'" I read Bradley's words aloud. "'I mean, if everyone sees only a monster and they keep treating him like a monster, how does he stop being a monster?'"

Ellie's eyes suddenly fluttered open. She was desperate to hear the counselor's answer.

The counselor tells Bradley that a person must first realize that they aren't a monster.

"'Until he knows he isn't a monster,'" I read the counselor's words aloud, "'how is anybody else supposed to know?'"

Ellie said nothing, but she heard everything. I paused long enough to listen to her all but imperceptible sigh.

What I wanted to say was: *You're not a monster! You're the best kid I know!*

Instead, I dog-eared the page, wrapped an arm around her shoulder, and said nothing.

"Sometimes people can learn a lot about each other," Bradley's counselor said, "just by sitting in silence."

*

At 9:00 a.m., following a good night's sleep, Ellie and I enter Fort Peck Interpretive Center, rested and ready for more dinosaurs. Maybe the change in scenery had changed our attitudes; whatever the reason, our dinosaur weariness from the previous day is now gone. At our tenth stop, we're recharged for the last leg of our journey—a mere four passport stamps away from the trail's completion.

Upon entering the interpretive center's lobby, we're greeted by Natural Resource Specialist Sue Dalbey, dressed in her ranger browns and badge.

"Well?" Sue asks moments after meeting us. "Did you get to see *Matilda*?"

"We did," Ellie confirms. "We loved it."

"We sure do enjoy the theater around here," Sue says.

"I just hope the dinosaurs can compete," I smile.

But judging by the life-size, full-skin skeleton cast of the *T. rex* prominently displayed in the lobby, the dinosaurs seem to be holding their own.

The interpretive center—a cooperative effort between the U.S. Army Corps of Engineers and the U.S. Fish and Wildlife Service—opened in 2005, the same year as the MDT. It likely wouldn't exist were it not for the 1997 discovery of Peck's Rex, which renewed the region's interest in *T. rex*.

"It was really a grassroots effort to build the Fort Peck Interpretive Center," Sue tells us as we peer up at the Peck's Rex replica glowering down at us. "We wanted to help people understand just how fossil-rich this area is."

Not that anyone should have needed reminding. From 1988 to 1997, two *T. rexes* were found around the Fort Peck Reservoir: first the Wankel *T.*

*rex* in 1988, followed by Peck's Rex nine years later. While the interpretive center features two versions of Peck's Rex (both the skinned version ahead of us and the cast version in the exhibition hall), the real version is displayed at Museum of the Rockies. Ellie and I had gazed upon it just nine days before while receiving our tour from Zach Perry. Peck's Rex was the first *T. rex* Ellie and I saw in Montana, and its dimensions (twelve feet tall and thirty-eight feet long) helped us understand the sheer magnitude of a species that defies imagination.

Peck's Rex is just one of Montana's many *T. rex* specimens. In a state spoiled with dinosaur discoveries, add twenty-four partial *Tyrannosaur* skeletons to the list. Montana is home not only to the most *T. rexes* but also to the first—discovered near Jordan in 1902 by renowned paleontologist Barnum Brown.

But just because Montana is *T. rex* country doesn't mean the specimens always stay in state. In 2014 the Wankels received word that the Wankel *T. rex* was leaving Bozeman's Museum of the Rockies—where it had previously been displayed—to serve as the "Nation's *T. rex*" at the Smithsonian Museum of Natural History. For years, the Smithsonian had struggled to find a *T. rex* specimen suitable to display in the nation's capital. In 2014, thanks to a fifty-year loan agreement with the U.S. Army Corps of Engineers (who owned the property on which the Wankel Rex was found), the Smithsonian finally had its centerpiece. When the Smithsonian reopened its newly remodeled fossil hall in 2019, museumgoers were treated to the Wankel *T. rex* as no one had ever seen it: skull pointed down, with its teeth locked into the frill of a *Triceratops* fossil. Excepting paleoartist Charles Knight's 1897 *Leaping Laelaps*, dinosaurs were rarely displayed in such a dramatic state.

While many Montanans were sad to see their *T. rex* go, they took solace in knowing that the Wankel *T. rex* had received the highest honor of being named the "Nation's *T. rex*." Its new home in Washington DC, assured that more than seven million annual visitors could now enjoy the specimen.

Yet where a *T. rex* goes, so go the tourists. While Montana could spare one *T. rex* for the good of the nation, it could hardly spare the tens of millions of dollars regularly generated by tourism. For several years, Sue Dalbey has served on the state's Missouri River Country tourism board, whose members are often confronted with how eastern Montana might compete with Montana's western half.

"How do we market eastern Montana to help people understand how beautiful it is and the kind of resources we have?" Sue asks. "It has its own unique beauty and enticement, but it's not what people think of when they hear the word 'Montana.' They think of the Rocky Mountains, Glacier, and Yellowstone. But it takes a specific effort to come here."

While eastern Montana thrives during the summer months, business slows precipitously during the other seasons.

"It's something we need to strategically keep in mind," Sue says. "How are we going to reach those markets with our unique experiences?"

*How many* T. rexes *is it going to take,* I wonder, *to persuade people to come to town?*

*

Ninety-three miles to the southwest, well beyond the reach of the reservoir, the town of Jordan rests comfortably in the prairies and rocky hillsides of the Hell Creek Formation. The town is home to 351 residents and an impressive number of dinosaurs—from the first discovered *T. rex* in 1902 to an *Edmontosaurus* found in 2023. In the time between, Jordan's Garfield County—and the Hell Creek Formation more broadly—has been ground zero for famous fossils, including Clayton Phipps's Dueling Dinosaurs, in addition to an array of theropods (think: *T. rex*), *Pachycephalosaurs* (dome-headed), and *Ankylosaurs* (the squat and impenetrable tanks).

About an hour outside of Jordan's Garfield County Museum—our eleventh stop on the trail—I call ahead to let the museum know we're headed their way.

The phone rings, but no one picks up—not altogether out of the ordinary for a local museum run mostly by volunteers.

Still, I check the website to confirm the hours.

"Uh-oh," I say.

"*Uh-oh* what?" Ellie asks.

"They don't open until 1:00 p.m."

"Is that bad?"

"Kind of," I say. "Because we'll be long gone by 1:00 p.m. At least if we intend to make Makoshika by nightfall."

This is, indeed, the plan: Fort Peck Interpretive Center, then Jordan's Garfield County Museum, then Glendive's Makoshika State Park, where we'll receive a guided tour with dinosaur YouTuber Dinodave Fuqua.

If it goes off without a hitch, it'll be another three-stamp day—leaving just two more stamps to go. But it appears we've hit our hitch. If we can't get access to the Garfield County Museum, so much for the completed passport.

I call the museum again. When that number doesn't work, I find a volunteer's home number buried in the bowels of the internet.

"Hello?" says the woman on the other end.

"Hi, there," I begin. "I'm sorry to bother you . . ."

I explain the situation, and the woman—summoning more grace than we deserve—provides me the home number of a fellow volunteer named Jacque, who has a key to the museum and lives nearby.

"I don't know if Jacque's available," the woman says. "Technically, the museum doesn't even open until 1:00 p.m. . . ."

My ask, I know, is enormous. And as needy, privileged tourists go, I suspect I'm at the top of the list.

Yet when I call Jacque, she's open—if not enthusiastic—about the prospect of helping us.

"We're from Wisconsin," I explain, "and I'm afraid we have another engagement later this afternoon. This is just terrible planning on my part. We were hoping to . . ."

"Who's 'we?'" Jacque cuts in.

"My daughter," I say. "Ellie. She's nine."

She considers it a while longer.

"All right," Jacque relents. "I'll meet you there in an hour."

This seems a perfect plan given the GPS's estimated arrival time. For the next twenty minutes, I'm so busy composing future thank-you letters to Jacque that I don't notice the traffic jam until I'm in it.

Perhaps "traffic jam" is too strong a phrase. What I mean to say is the enormous truck with the "oversized load" sign that appears to be hauling some space-age contraption of indeterminate origin and purpose. The truck is accompanied by a pilot car, whose flashing lights ensure I don't get close.

Suddenly, our seventy-mile-per-hour open road crawls to a near standstill.

The two-lane road offers us few options. Either we wait or . . . we wait.

To make matters worse, we seem to have hit the only road in Montana currently under construction. The road turns to dust, clouding the air and creating a smoke screen that blots the world beyond the windshield.

On the GPS screen, our estimated time of arrival increases by several minutes. I imagine Jacque pulling into the Garfield County Museum lot and wondering where we are. Maybe she checks her watch. Maybe she tells herself she'll give us a few more minutes, but then she's gone.

"Come on," I mumble, drumming the steering wheel. "Please let us go around . . ."

"Yeah, come on, you dumb truck!" Ellie echoes.

The truck and pilot car make no attempt to accommodate our requests.

Help, thank goodness, comes from behind—a small caravan of pickups that strengthens our case for letting us pass.

"Look in your rearview," I shout to the pilot car.

"Just look already!" Ellie hollers.

At last, the oversized hauler and pilot car pull to the side of the road. Since I haven't seen any oncoming traffic for the past half hour, I turn sharply toward the opposite side of the road, into the dust, passing as fast as I can.

After four seconds of slight terror, we emerge from the dust cloud no worse for wear.

"Yes!" I shout.

"Heck yes!" Ellie shouts.

Eyes wide, pedal to the metal, this is our kind of country road.

*

Ellie and I race across the Garfield County Museum parking lot to find the door unlocked.

In the gift shop, a woman I presume to be Jacque Gregg looks up from her paperback book.

"Oh, Jacque," I say, "I'm so sorry to have dragged you out here."

"Well," she says, "that's why my name is on the door."

(It is, scribbled on a piece of paper along with her home phone number.)

"You do this kind of thing a lot?" I ask.

"Sometimes," she says. "I came in for you because you have your daughter with you. Kids need to see history like this."

"You're very kind," I say.

"It's nothing," Jacque says, turning toward Ellie. "And I'll stamp your passport for you if you like."

Gratefully, Ellie hands it over.

"I've got to find my stamp," Jacque says, scanning the gift shop. "In the meantime, why don't you two have a look around? I already turned the lights on for you."

Ellie and I enter the museum's main corridor, where we find shelves and open spaces full of artifacts from the region's past. From homesteading history to dinosaur fossils and casts, the room sprawls on for what seems like miles. To our right, we see one of the museum's crown jewels.

Not a dinosaur, but a hammer: the one used by Barnum Brown as he unearthed the world's first identified *T. rex* in 1902—not far from where we stand.

*

Born amid the Bone Wars in 1873, Barnum Brown was perhaps destined to eclipse his warring paleontological predecessors, Edward Drinker Cope and Othniel Charles Marsh. Regarded by many to be the "greatest dinosaur collector of all time," he was also known by a less soaring title: Mr. Bones. Throughout his life, Brown adopted a few eccentricities (including free-lancing as an intelligence asset during World Wars I and II and being the only paleontologist to don a fur coat on a dig site). Yet these details seemed fitting given his namesake: showman P. T. Barnum.

Like paleontologist Charles Sternberg, Brown's humble beginnings might have kept him from the dig sites if not for the patronage of a man who shared his passion for all things prehistoric. Henry Fairfield Osborn, the head of the American Museum of Natural History's Vertebrate Paleontology Department, was for Brown what Edward Drinker Cope was for Sternberg: a ticket to the dinosaurs of the West. In some respects, Brown was an amateur (taking a leave from college to join the digs), but he quickly catapulted to become one of the most accomplished people in his field. Like Cope and Sternberg, Barnum Brown was compelled to get his hands dirty.

"When faced with the choice of either staying home to complete his research or escaping the confines of the museum," Brown's biographers wrote, "he almost always chose to pursue his opportunities in the field."

In the summer of 1896, twenty-three-year-old Brown was recruited by a paleontologist named Jacob Wortman to assist on a dig. Brown proved so valuable that Wortman recommended him to Osborn, who soon promoted him to assistant curator.

Henry Fairfield Osborn could more regularly be found in the eastern museums than the western dig sites, meaning that much of the actual paleontological work was left to Brown, who both disappointed and pleased his boss. In July 1902, shortly after receiving some of Brown's recently discovered fossils, Osborn chastised his protégé for being too hasty in shipping fossils of little scientific value.

"This seems to warn us that we should certainly examine material a little more carefully in the field," Osborn wrote, "before taking it up in a block and going to the heavy expense of freight shipment to the east."

Brown accepted the chiding gracefully and, in his reply to Osborn, mentioned a more recent fossil find—destined to become one of the greatest of the twentieth century.

Brown described the fossils in detail, noting that he believed they belonged to a previously unknown carnivorous dinosaur.

"I have never seen anything like it from the Cretaceous," he added.

With such a find, Osborn quickly forgave the young Barnum Brown.

When Brown and his crew returned to the site in 1905, they continued their difficult work. They "attacked the hillside with plow and scraper," Brown wrote, "but soon the sand was so hard that the plow was no longer effective so we sent to Miles City for dynamite to drill and blow off sections of the hillside."

Their hard work paid off, introducing the world to a giant carnivore unlike any it had ever seen. *Tyrannosaurus rex* was scientifically named in 1905 and appeared—much to the delight of readers—in a widely-publicized article in *The New York Times* in December of that year.

The article hailed the newly discovered species as "the most formidable fighting animal of which there is any record whatever," adding that the *T. rex* was "the war lord of the earth in his day." Barnum Brown was fully credited with the find, the newspaper noting how he had spent years in the Montana badlands—"perhaps the most inaccessible spot in America" according to Brown—yet was rewarded with a find that would surely go down in the annals of paleontological history.

Yet following Barnum Brown's death in February 1963, *The New York Times*—the same paper that once celebrated him—ran a short obituary that (for reasons I'll never understand), curiously omitted his *Tyrannosaurus rex* discovery. Though the piece hailed Barnum as the "Father of the Dinosaurs," there was no mention of the specific dinosaurs discovered during his seventy or so years of paleontological service, including his most important discovery: the unearthing of the world's first identified *T. rex* skeleton.

Today, Barnum Brown's name is synonymous with his most famous find. Call him what you will—"Mr. Bones," "greatest dinosaur collector of all time," "Father of the Dinosaurs," "the guy in the fur coat"—they all translate to *Tyrannosaurus rex*.

*

Just beyond Barnum Brown's hammer in the Garfield County Museum, Ellie and I spot a display dedicated to the Wankel family and their own *T. rex* discovery nearly a century later. I can't help but admire Barnum Brown and the Wankels' contrasting stories, which collectively confirm that almost anyone has the potential to find a dinosaur. While Barnum Brown dug long and hard to discover his, the Wankels were fortunate enough to have picked the right campsite during drought season.

Ellie and I linger at the far end of the museum—past a full-size *Triceratops* cast—until my eyes fall to a less dramatic artifact: this one from my own deep past.

"I can't believe this . . ." I whisper, walking swiftly toward a poster affixed to the museum's far wall.

Ahead of me is something I haven't seen in three decades, though I saw it daily for a time: a full-color print featuring the most famous dinosaur species of the bunch. The poster was a mainstay of my childhood bedroom, where I'd marvel at it late at night as the streetlights glowed through the window shades.

Back then, dinosaurs represented something unfathomable. More fairy tale than fact, yet they were real. Seeing that poster again—for the first time in thirty years—takes me back to that bedroom wall, and that moment in my past when my "dinosaur dream" involved getting to know them better. For a time, all I wanted was to whisper their names as if summoning them back to life.

*Triceratops. Stegosaurus. T. rex . . .*

It feels good to reunite with my old pals. My very old pals, indeed.

Back at the gift shop, Jacque asks Ellie, "What'd you think?"

"It was awesome," Ellie confirms.

"Are the dinosaurs what interest you most?" I ask Jacque.

"Oh, I'm more interested in homestead genealogy," she says.

Jacque's rancher-farmer grandfather found his way to Garfield County in 1911, less than a decade after Barnum Brown's discovery of *T. rex*. Two generations later, Jacque and her husband, like her grandfather who came before, also made their livelihoods from working the land, as do most residents of Garfield County.

"But these days," Jacque tells us, "it's hard to make a living that way."

FIG. 21. The author, at age nine, posing with his new dinosaur poster, which he rediscovered thirty years later at the Garfield County Museum in Jordan, Montana. Author photo.

It's the same story in Choteau, Bynum, and Rudyard.

But Jordan seems particularly hard hit. Over the past sixty years, the town's population has decreased by over 40 percent.

"Most folks are selling their ranches and moving elsewhere," Jacque reports. "My husband sold our ranch . . . I guess it'll be two years this September."

"Was that a hard decision?" I ask.

"It wasn't," Jacque says. "His health was getting bad. He passed away about six months ago."

"I'm sorry."

"He was hanging on for our anniversary," Jacque says, her voice softening. "It was going to be our fiftieth. But the Friday night before he passed away, he told me he wasn't going to make it. He was feeling kind of blue, but he was ready. And he had all the kids back home for almost a week, so that made him happy."

I glance toward Ellie.

"Not all children come back," Jacque warns me, handing me our freshly stamped passport. "It depends on how you raise 'em."

*

What better way to love our children than by introducing them to the world of dinosaurs? That's been my theory, at least. And it's a theory that parents have been trying for nearly a century—long before there was ever a Montana Dinosaur Trail.

Dinosaur tourism may have started in 1852, just a decade after British anatomist and paleontologist Sir Richard Owen coined the term *Dinosauria*. That year, British artist Benjamin Waterhouse Hawkins was commissioned to create a menagerie of clay and cement dinosaurs for London's Crystal Palace Park. The result was a series of reptilian creatures (considered entirely inaccurate by today's standards) that nevertheless sparked the world's imagination by offering an artistic representation of the creatures whose bones had been unearthed. The Crystal Palace dinosaurs (as they became known) became a nineteenth-century marvel: a glimpse into a deep past that humans could hardly fathom. British children gawked at the wonders, and soon American children would too.

Or they almost would.

In 1868 Hawkins brought his dinosaur-building talents to New York, where a Paleozoic Museum was proposed for Central Park. The plan fell through, delaying the majority of large-scale artistically rendered dinosaurs in the United States for seventy years.

It wasn't until 1936—amid the Depression—that South Dakota's Rapid City Chamber of Commerce saw an opportunity to create jobs and boost tourism on the government's New Deal dime. While the Works Progress Administration was supporting the construction of Fort Peck and its dam, four hundred miles away in Rapid City, five concrete dinosaurs were erected through the same program just off U.S. 90. For nearly ninety years, Dinosaur Park has served as a kitschy roadside attraction for travelers en route to Mount Rushmore. Generations of road-trippers have posed alongside the

FIG. 22. Artist Benjamin Waterhouse Hawkins's "dinosaur studio" in England, where he created dinosaur models for Crystal Palace Park. Wikimedia Commons, https://en.wikipedia.org/wiki/Benjamin_Waterhouse_Hawkins#/media/File:Sydenham_studio.jpg.

FIG. 23. A mid-twentieth-century family posing alongside a sauropod in South Dakota's Dinosaur Park. Sternberg, George Fryer 1883–1969, "064_03: Group Photo in Front of a Dinosaur Model" (2021). *George Sternberg Album #4*. 295, Forsyth Library Special Collections, Fort Hays State University.

park's giant dinos, which provided, if not an entirely accurate representation of dinosaurs, a fun one, nonetheless.

Despite Dinosaur Park and similar roadside attractions, it wasn't until 1964 that dinosaurs fully reentered the American consciousness. Following a fallow dinosaur period at the end of the Bone Wars, the creatures came roaring back at the New York World's Fair. Sinclair Oil's Dinoland was a fair highlight, introducing more than fifty million Americans to nine fiberglass dinosaur replicas.

Wildlife artist Louis Paul Jonas constructed the dinosaurs under the guidance of paleontologist Barnum Brown, who, at eighty-nine years old—and more than sixty years removed from his *T. rex* discovery—was still actively engaged in educating the public on dinosaurs. This wasn't Brown's first consulting gig; he'd also worked on *Fantasia,* bringing a sense of realism to the dinosaurs depicted in "The Rite of Spring." Barnum Brown did for

FIG. 24. A mid-twentieth-century family posing alongside a theropod in South Dakota's Dinosaur Park. Sternberg, George Fryer 1883–1969, "065_05: Group Photo in Front of a Dinosaur Model" (2021). *George Sternberg Album #4*. 303, Forsyth Library Special Collections, Fort Hays State University.

*Fantasia* what Jack Horner did for *Jurassic Park*: ushered more scientifically accurate dinosaurs to the big screen.

While Barnum Brown's contributions to paleontology are numerous, perhaps his most often overlooked role is his service as a bridge between two eras. Born in 1873, the golden age of dinosaurs, Brown lived until 1963, the beginning of the Dinosaur Renaissance. In a single lifetime, he witnessed one of the most dramatic shifts in our understanding of dinosaurs.

As paleontologist Robert Bakker wrote in 1975, dinosaurs were once "symbols of obsolescence and hulking inefficiency," but after a century of study, our more modern understanding viewed them as "immensely more sophisticated" than previously thought. Gone were the cold-blooded reptiles; enter the warm-blooded creatures more closely akin to birds.

Although Brown died shortly before this dramatic shift gained scientific acceptance, his lifetime of work helped usher in a new wave of study. Perhaps more than most, Barnum Brown inspired a love for dinosaurs throughout the twentieth century. He was dinosaurs' greatest champion, an early example of a science communicator who took his knowledge beyond scientific journals to reach everyday people—most notably through his weekly CBS radio show in the latter years of his life.

Science benefited from Barnum Brown's discoveries and outreach but so did the public.

Not to mention dinosaur tourism in all its forms—from the big screen to the roadside attraction.

Before the rise of the *Jurassic Park* generation of paleontologists, there was the 1964 World's Fair generation. And before them were those inspired by places like Rapid City's Dinosaur Park.

Barnum Brown's contributions didn't end with discovering a *T. rex* near Jordan in 1902. Those bones were just his beginning.

*

We head east on Montana US-200, past Brockway and Circle, until arriving at the day's destination of Glendive. With a population just shy of five thousand, Glendive's a metropolis compared to where we've recently

FIG. 25. Ellie and YouTuber Dinodave Fuqua exploring Makoshika State Park in Glendive, Montana. Author photo.

been. It's home to a host of dinosaur-related experiences, too, including two Montana Dinosaur Trail stamps, the Frontier Gateway Museum (where we briefly stopped for our twelfth stamp), and Makoshika State Park (where we received lucky 13). But Makoshika State Park has far more to offer than a stamp. It's also the preferred stomping ground for YouTube sensation Dinodave Fuqua.

Ellie and I first stumbled upon Dinodave eight months before while watching YouTube videos from every town we hoped to visit. Upon clicking on our first Dinodave video, we were smitten by the man's high-octane enthusiasm for all things dinosaurs.

"Today, we are going to go on the Diane Gabriel Trail," the bandanna-headed Dinodave told the camera, "and I'm going to show you some things on this trail that nobody knows about!"

His passion was palpable, and by week's end, I'd reached out, and he'd agreed to take us on a guided tour.

No sooner do we arrive at Makoshika State Park's visitor center than Dinodave circles into the parking lot.

"Hey guys!" he shouts from his car window. "You ready to hike or what?"

The correct answer is "Yes!" though since the temperature is teetering dangerously close to triple digits, our "Yes" lacks enthusiasm.

Thankfully, Dinodave has enough enthusiasm for us all.

We park at the trailhead for the Diane Gabriel Trail, where Dave bends to tie his shoes.

"You guys got enough water and stuff?" Dave asks.

"I think we're all set," I say. "Right, Ellie?"

Glancing up at the steep spires and hillsides abutting the sky, Ellie gives me her *Not-set-at-all* look. But by now, Dave is already tromping down the trail; we have no choice but to follow.

The landscape is like nothing we've ever seen. The park's eleven thousand acres make up some of the only badland formations in the United States. While the Lakota translation of Makoshika means "bad land," when Europeans set foot upon centuries later, they offered a different description: hell without the fire.

Yet Makoshika isn't just physically hot; it's also a dinosaur hot spot. Over the past century, fossils from at least ten dinosaur species have been discovered on this land.

The sun soaks into the rock but also into our bodies. It beats upon us like a heat ray, zapping our energy. Still, we trudge forward, tromping mountain goat style up the hillside as broken bits of sedimentary rock slip beneath our feet. I pause for a breath, peering out at the desolate wonderland littered with slopes, gullies, and the park's most unique geological feature, hoodoos—spires of rock worn thin by erosion, occasionally topped by a mushroom shape. Every cutback reveals some new novelty, but for Dinodave, it's just another walk in the park.

Before he was Dinodave, he was simply Dave Fuqua, a boy born and raised in Glendive.

"We'd come out here all the time as kids," Dave says. "It was just wild and free. My dad took us out here and showed us what dinosaur bones were, and I just thought, 'Why are they here?' It just didn't make sense to me." His childlike curiosity drove him to learn more. Dave's family moved around during his high school years, but he returned to town to attend classes at Dawson Community College.

"I didn't even take any geology course," Dave says. "I just read as many books as I could. In my spare time, I'd go hiking, and the more I hiked, the more I understood the rocks."

In 1991 Dave learned that Museum of the Rockies paleontologist Diane Gabriel was digging up a *Triceratops* skull near Makoshika State Park. While amateur paleontologists had to pay to assist in the dig, Dave—an on-again, off-again college student—was in no position to do so.

"It was a secret spot," Dave says. "But I found it."

One afternoon, eighteen-year-old Dave rolled up to the dig site on his motorcycle, fully expecting to be turned away. Instead, Diane welcomed him and shared what she'd been digging.

Dave was so grateful for the hospitality of the professional paleontologist that he offered to share a few of his own secret dinosaur hotspots with Diane.

Diane agreed to have a look.

Over the next few months, Dave and Diane explored various dig sites together, and Dave peppered Diane with as many questions as he could.

"She kind of tutored me," he explains.

Recognizing Dave's passion, Diane encouraged him to enroll at Montana State University, where she could also mentor him in the classroom. Though Dave never considered himself a "student type," he was willing to try in exchange for the opportunity to work with Diane.

Tragically, Diane was diagnosed with breast cancer soon after, and she died in 1994.

Dave was devastated and, without his mentor's support, left dinosaurs in his past.

At least, he thought he had. But six years ago, Dave and his buddy made a short video about their dinosaur adventures in Makoshika State Park. Unbeknownst to Dave, his friend edited the video and uploaded it to social media, then YouTube.

Ten thousand views later, Dinodave was born.

"He didn't tell me he was going to do this," Dave explains. "We were just kind of screwing around. But then it went viral. And he proved to me that I *could* do this."

The timing couldn't have been better. At forty-six, Dave had entered his self-described "mid-life crisis."

"At the time, I had nothing going for me," Dave says. "None of my dreams had panned out." Dinodave Paleo Adventures changed everything, heightening his profile and renewing his love for all dino-related things.

In recent years, Dave's fossil finds in and around Makoshika State Park have garnered interest from Museum of the Rockies on down. But his most recent find—a discovery with the potential to be a new dinosaur species—excites him most.

"It'll take a few years for the scientists to write it up," Dave says. He's been working with paleontologists at the Burke Museum of Natural History and Culture. "And they want to go back to the site to make sure they have everything first. But I mean . . . finding a new species would be unbelievable."

Dave says his relationship with the Seattle-based Burke Museum has been overwhelmingly positive. And it further proves the potential for partnerships between amateur paleontologists like Dave and professional paleontologists like those at the museum.

"So if you did discover a new species," I say, "is there some financial upside for you?"

"Zero," Dave says, "because it was found on government land."

Dave knows the law as well as anyone. The basic rule is that fossils found on public land belong to the public, as stipulated in the Paleontological Resources Preservation Act of 2009.

While he's floored by the enormous paydays received by commercial fossil hunters (who find fossils on private land and cut a deal with the landowner), that work isn't for him.

"I am broke," Dave says, "so I'm banking on this YouTube thing working out. I'm just trying to go about this the right way."

The truth is, discovering dinosaur fossils is only lucrative if you're willing to ignore the Society of Vertebrate Paleontology's code of ethics regarding fossil sales. Folks like Dave would love nothing more than to make a living finding fossils, but selling fossils—even those legally obtained on private land—is a red line he's unwilling to cross.

"But if I found a million-dollar [fossil] on someone's private land, would I sell my soul?" Dave asks aloud. "Well, that's a lot of money."

Dave receives no finder's fees for bringing his discoveries to the attention of professional paleontologists. Even recognition can be hard to come by. But for Dave, that's beside the point.

"I just love looking for dinosaur bones. That's what I like," he says. "Whatever it takes, I just want to do that."

Still, I can't help but imagine the alternative history: the one where Diane Gabriel had lived and Dave had benefited more fully from her tutelage.

With a few more classes and degrees, might Dinodave have been Dr. Dave? Might he have joined other professional paleontologists in shaping a future from the past? And been able to pay his bills while doing it?

FIG. 26. Ellie among the beauty of Makoshika State Park. Author photo.

*

That evening, as the temperatures fall, Ellie and I pull on sweatshirts and take one last walk along Makoshika's main road. Our tent is one of a handful in the camping area, where our neighbors sprawl in camp chairs, books in hands, occasionally glancing up from their pages to take in the view.

It is a wonder to behold. In the foreground, the sagebrush abuts the sandstone hillsides, whose layers are defined by a rainbow of shades. Overhead, waves of cumulus clouds clot the state's famous big sky—a rotation of pink, purple, and blue, depending on the time of night.

As the last vestige of sunlight dips beneath the hills, Ellie and I sit on a secluded bench a quarter mile from camp. After nearly two weeks on the road, we are running short on things to talk about. We can only say what we always say these days: "I can't believe that we're here."

A few minutes before 9 p.m., night casts its long shadows from the rocks, slowly coating us in darkness. It's been a long day—from Fort Peck to Jordan to here. We are exhausted in the best way; our minds and bodies are beat.

Back in the tent, I try to read aloud from *Ramona and Her Father*, but I barely make it two pages before Ellie drifts to sleep. Somehow, the burning temperature has given way to a chill, so we cuddle close for warmth. Outside the tent's mesh window, the moon hangs like a prop from the sky. Up the hillside, a light wind kicks up dust.

Maybe we'll rise in the morning to find some dinosaur fossils newly exposed in the dirt. Who can say?

Sometimes, amateurs get lucky. It wouldn't be the first time.

FIG. 27. Ellie and Glendive Dinosaur and Fossil Museum director Robert Canen. Author photo.

# 11
# Of Faith and Fossils

### JULY 10
*Glendive, MT*

Shortly after sunrise, I wake to the western meadowlark. The bird works through the scale like a trained soprano, easing me from the dream world. Yawning, I unzip the tent and walk toward the picnic table, a book in tow. But I barely crack it before a sound off to my right startles me. I glance up to spot a man in his sixties lumbering toward me, a canister-shaped contraption in his hand.

"Hey, buddy," the man whispers, settling across from me at the picnic table, "you ever seen one of these?"

I eye the contraption. "Can't say I have," I say.

"Portable coffee maker," he says proudly. "You just fire up the gas, press this black button here, and . . ."

At the sound of the click, a small flame rises from a miniature fuel container.

"When this goes orange," he continues, pointing to an indicator light, "that's when you know it's done."

"That's pretty slick."

"It'll give you sixteen ounces of coffee," he says. Then, in the way of introduction, he adds, "James Kelly."

"B.J.," I say, giving him a nod from the opposite side of the picnic table.

I'd first spotted him late the previous evening. He'd rolled into a campsite on his Harley-Davidson, retiring to his tent shortly after the stars came out. Upon getting a closer look at him by morning light, I see that James, indeed, looks the part of a biker—strong, sturdy, and with two-day-old stubble coating the lower half of his face.

As the water heats, James and I chat. He's sixty-one and an employee at the Hanford Site in eastern Washington—a decommissioned nuclear production area that, back in the 1940s, was home to the first full-scale plutonium production reactor on the planet.

("It's where they made the plutonium for the Nagasaki bomb," he tells me.)

But for the next week and a half, he's leaving all that behind. He's two days and nine hundred miles into a roundtrip adventure to a Harley-Davidson convention in Milwaukee, four hours southeast of our home in Eau Claire.

The only thing James enjoys discussing more than coffee is his motorcycle, a 2023 touring bike complete with saddlebags. We talk about his history of bikes and where they've taken him—forty-eight states in all. He describes a five-week road trip from several years back during which he rode through thirty-six states.

"Eleven thousand miles total," he says, pouring the hot water into his press. "I went 620 miles in one day." He rattles off his itinerary as if reciting the ABCs. "Oh, and you gotta check out these prairie dogs I saw out there," he says, turning his attention from the press to the photos on his phone. "Look at 'em standing next to each other," he chuckles, pointing his phone my way. "They look like they're getting married."

There's something about an early morning chat with a stranger—discussing prairie dog weddings, no less—that awakens the body and soul at least as well as coffee. For the next fifteen minutes, we talk about family, work, travel, books, and dreams. I tell him about dinosaurs, the Montana Dinosaur Trail, and the circumstances that have led us to this place.

"So you're gonna write a book about it?" he asks, lifting his freshly pressed coffee to his lips. "About dinosaurs and your daughter and all that?"

"I'm going to try," I say. "But it's also about the towns we visit. And the people we meet along the way. Like you," I add.

"Me, huh?" he says skeptically.

"Sure," I say. "You've got a great story. A guy driving around the country on his motorcycle, seeing it all. I'd love to write about that. If you don't mind."

He mulls it over.

"I don't mind," he says at last. "But you've got to say I'm 'riding' around the country, not 'driving.' You 'ride' a motorcycle. You don't 'drive' them."

"Noted," I smile.

James takes a long sip of coffee, his eyes scanning the light warming the rocks just ahead.

"The thing about riding a bike is that it lets you work through all your issues," James says. "You forget you got a job at home, or anyone who depends on you for anything. The problems kinda just . . . go away."

"That's what I love about road trips too," I admit.

"And I put on ten times as many miles as most people," James continues, nodding toward his bike. "Because I'm free."

"Free how?"

"I'm an alcoholic," he says. "But I haven't had a drink in twenty-nine years. On the nineteenth of June, I celebrated that. And because I'm not recovering from something, I can roll anytime I want. I'm never hungover; I never feel bad. I just go."

I think of my former student and Earth Sciences Foundation director Tom Hebert, who recently celebrated twenty years of sobriety and, like James, now prioritizes dreams over drinks.

Except for coffee, in James's case.

I turn my attention to the whir of the tent unzipping, then spot Ellie—as wobbly-kneed as a newborn fawn—stumbling barefoot toward us across the grass.

"Want some coffee?" James offers, lifting the container.

"Um . . ." Ellie gives me her *who-the-hell-is-this* look. "Not really."

"This is James," I say.

"Hi," she yawns.

"You know what you need?" James tells her. "Some morning stretches. I need 'em, too. Here. Let me show you."

A moment later, James leads Ellie through a private yoga class in the shadows of the spires. His arms spin like whirligigs, his torso rotating like a kitchen mixer. Then, with a grunt, James leans into a downward dog.

"And this one is the sun stretch," he informs her, rising from the downward dog and entering mountain pose. He clasps his hands and directs his face toward the sun.

Ellie is a perfect student, matching him move for move. And when it's over—once the biker has completed his stretching routine alongside my daughter—he laughs and waves goodbye.

"Now you can go back home and show all your friends," he chuckles. "Tell 'em I taught you the sun stretch."

The meadowlark leaves its post and flies away. Soon after, James's motorcycle thrums off in the same direction.

"So . . . who was that again?" Ellie asks as we roll up the tent.

"James," I say matter-of-factly. "Some guy named James."

*

Next door to Glendive's Frontier Museum—where we'd received our stamp the previous day—is another museum, the Glendive Dinosaur and Fossil Museum. Given their literal side-by-side placement (they're separated only by a few hundred feet), I'd conflated the two in my head. I shouldn't have.

While the Montana Dinosaur Trail–approved Frontier Museum showcases dinosaurs from a traditional scientific perspective (read: that non-avian dinosaurs went extinct sixty-six million years ago), the Glendive Dinosaur and Fossil Museum tells a different story with a different timeline.

Founded in 2009 and owned by the nonprofit Foundation Advancing Creation Truth (FACT), the Glendive Dinosaur and Fossil Museum offers a creationist perspective on the planet's age and formation.

Meaning: the museum espouses the belief that Earth was created by God in seven days and is somewhere between six thousand and ten thousand years old. This starkly contrasts my personal view and, more importantly, the more scientifically held perspective that Earth is the result of 4.5 billion years of changes over time. This isn't exactly a "split the difference" situation. Nor is it easy to find a middle ground when the evidence comes from such vastly different sources. The older date is a science-based fact determined by radiometric dating; the other is a faith-based belief based on a literal scripture reading.

Young Earth creationism (as it's often called) seems like a fringe idea. However, a 2019 Gallup poll revealed that 40 percent of U.S. adults believe God created humans in their current form within the past ten thousand or so years. Protestant churchgoers without a college degree were the most likely to believe in creationism. Those who didn't attend church and had a college degree tended to take a more evolutionary approach.

We're not just talking about the age of Earth, but its many implications, including how, if the planet is six thousand years old, dinosaurs could have gone extinct sixty-six million years ago.

Robert Canen, the museum's director, has answers. I'd first learned of Robert and the Glendive Dinosaur and Fossil Museum from Tom Hebert.

"It's a version of the story you won't hear in any other Montana museums," Tom assured.

In the spirit of "thoroughness" on all things dinosaurs—and perhaps to test my ability to listen to ideas well beyond my own—I emailed Robert to try to set up a last-minute interview. In my request, I'd clarified that I was working on a book and did not subscribe to the creationist narrative.

I expected silence, or at least some "thanks but no thanks" that might spare us both from an uncomfortable conversation. Instead, Robert responded within hours.

*Hi, B.J.! We would love to meet with you on Monday, July 10 for an interview. I could definitely walk with you around the museum and answer any questions you might have. . . . I will put it on my calendar. Let me know if anything changes.*

*Have a great day!*

I was perplexed by his reply. Had I not clarified my intention to publish a book espousing my completely contradictory view? Did he not worry that some out-of-town academic like me might turn him into a punchline?

Or perhaps I was the one being played—a conduit capable of sharing his message with a broader audience.

*Which one of us is walking into the lion's den?* I wondered.

*

"Sorry I'm late," Robert says, striding toward us in the museum lobby.

(He's not late; in fact, he came into work on his day off to meet us.)

Dressed in jeans and his "Grandpasaurus Rex" shirt, fifty-three-year-old Robert—bespectacled and bearded—greets us with a smile.

"You must be Ellie," Robert says. "I see you're already working on our scavenger hunt."

"Yup," Ellie agrees, relieved to have been upgraded from activity sheets to scavenger hunt. "I'm counting clownfish."

"Clownfish, huh?" Robert says. "Well, let's have a look together."

We enter the exhibition hall, which features some of the most impressive dinosaur displays since Museum of the Rockies. Murals coat the walls, while fully articulated dinosaurs surround us. Much of the museum's first-floor signage offers the usual fare—information identifying the dinosaur species and a few words about its discovery. However, some of that signage also raises questions about the fossil record and timelines, furthering the creationist perspective.

Most museumgoers know what they're in for. Indeed, the museum makes no secret of its intentions. Next to their "No Food or Drink" sign is another sign clarifying the museum's commitment to displaying its exhibits "in the context of biblical history."

Some museumgoers pilgrimage to the Glendive Dinosaur and Fossil Museum to reaffirm their faith and hear the creationist version of history, though others come simply to see their vast collection of dinosaur fossils and casts.

"You don't have to believe what we believe to enjoy your time here," Robert assures. "We're just glad you came."

This is precisely my plan. I will listen and enjoy my time with the dinosaur fossils. What I will not do, I suspect, is change my mind about the age of the earth.

In the rhetoric and composition classroom where I taught Tom Hebert, I regularly referenced digital creator Dylan Marron's idea that "empathy is not an endorsement." We can listen to a person, even care about that person, without needing to endorse their ideas. This is the tack I choose to take with Robert.

Within moments, we're talking about timelines.

"But doesn't the age of rock layers alone dispute the creationist timeline?" I ask.

"It all depends on how you date the layers," Robert counters. "While an evolutionist would say the layers represent millions of years, we would say the layers generally represent the year of Noah's flood."

Mainstream geologists are quick to disagree. Their radiometric dating reveals that the rock layers far transcend the creationist.

"So, according to your timeline," I begin, "humans and dinosaurs lived simultaneously?"

"Correct," Robert says.

Mainstream paleontologists disagree. Their radiometric dating shows that dinosaur fossils are at least sixty-six million years old.

"And then the flood came and washed everything away?"

"Yes."

Again, mainstream scientists disagree, in part because there's not enough water for a global flood. According to the U.S. Geological Survey, if all the water in the atmosphere deluged Earth at once, the world would be covered to a depth of just under an inch.

I measure my next words carefully, an honest question that may prove alienating.

"So . . . how did Noah fit all those dinosaurs into the ark?"

Robert smiles. I suspect he's gotten this one before.

"Look at cats," he says. "A lion has the genetics to breed with a house cat. There's a size issue, but if you take two cats with all the genetic info to make lions and house cats, you end up with a lot of [species] variation."

By some measures, he's right. A 2013 study in *Nature* confirmed that house cats do, indeed, share 95.6 percent of their DNA with Amur tigers. His point is that even a limited number of the ark's animal inhabitants could have potentially paved the way for a wide variety of diverse species. Thus, the ark may have held more house cats than lions (as well as the dinosaur equivalent).

Regarding evolution, Robert says, "We believe in change. We don't think that species are static and can never change. We just think there are boundaries to those changes. We don't look at it as an evolutionary tree. We look at it as an orchard."

The tree-versus-orchard explanation is another favorite in creationist circles. Creationists believe that species emerge from an orchard's worth of various trees of animals—cats, dogs, elephants, you name it—and that a variety of related animals springs forth from each trunk. Evolutionists, in contrast, believe *all* species spring forth from various branches of the *same* tree.

While Robert and I disagree on the tree-versus-orchard debate, he'll later remind me that our differing perspectives—on this and related topics—might boil down to where we place our faith in the scientific realm. If science is "the systematic study of the structure and behavior of the physical and natural world through observation, experimentation, and the testing of theories against the evidence obtained" (Robert's definition, and one not far off from the *Oxford English Dictionary*'s), then we ought not to overlook the "observation" part, Robert implies. Later, he'll compare it to a courtroom, noting that various versions of events are often cited as evidence—from eyewitness accounts (observation) to more scientific-based evidence. Collectively, both types of information (eyewitness accounts and scientific-based evidence) provide more than either independently.

"Scientific inquiry can help reconstruct history," Robert says, "but it definitely cannot do it on its own."

Robert, who earned his master's degree in apologetics from the Institute for Creation Research, has an answer for every question I ask. Indeed, I ask a lot of them, many of which follow the formula, "But how can you believe X if science tells us Y?"

I don't mean to try to poke holes in Robert's beliefs; I'm here to understand where he's coming from. But where he's coming from is a place I'm only visiting and where a brief visit seems long enough.

We are not just talking about the age of Earth. Tied to it is a larger question regarding the role of science in the modern world. A 2015 Pew Research poll revealed that only 30 percent of Americans say their religious beliefs conflict with science. Of those, however, 36 percent believe that conflict centers on differences in the "creation of the universe, evolution, and Darwin." Twenty-four percent of those who note a conflict attribute it to "broad differences over the belief in God, facts v. beliefs, miracles, view of man as 'in charge.'" Eleven percent note a conflict in "views about the beginning of life, abortion." Seven percent note a conflict related to "specific medical practices."

Thus, a creationist's disagreement with the radiometrically dated age of the earth is but an entry point into science and religion's other "conflict areas." Many of those revolve around the most heated issues of the day, from abortion to in vitro fertilization to stem cell therapy.

I could fill an ark with all the differences between Robert and me. And yet I could fill a second ark with everything we've got in common. Sure, we're 4.5 billion years apart on the age of the earth, but we're both fathers and writers who share a love for dinosaurs. And come Sunday mornings, we'll both find ourselves seated in pews. Albeit different pews in different churches—one espouses a strictly literal interpretation of the Bible, while the other prefers a more metaphorical approach to scripture.

Despite these differences, Robert and I agree that science and faith are best approached humbly.

"I have no problem admitting when I'm wrong," Robert remarked in a 2016 article in the *Great Falls Tribune*, "and that's something science and this whole field could use more of."

Of course, the best science is quick to admit when it's wrong. After all, the long road to scientific truth is littered with incorrect hypotheses, not to mention theories that are repeatedly revised as new information becomes available. Are scientists sometimes slow to acknowledge errors or incorrect theories? Perhaps. But that's as much a human problem as a scientific one.

Since its founding in 2009, more than a few professional paleontologists have condemned the Glendive Dinosaur and Fossil Museum, calling it a "temple of ignorance" and "the opposite of a science museum."

One of the most compelling critiques came from paleontologist Mary Schweitzer, who famously discovered *T. rex* soft tissue in 1983. For the first time in history, scientists could determine the dinosaur's sex based on the tissue's cellular makeup. However, the discovery also emboldened creationists to try to employ the newly discovered tissue to further their "Young Earth" timelines. Creationists claim that the degradation of the tissue supports their position.

"I'm a very strong Christian," Schweitzer said in 2009. "My faith means everything to me." Yet she's wholly opposed to creationists co-opting her discovery to try to confirm their timeline. As she notes, not nearly enough is known about tissue degradation to make such a claim. For Schweitzer, the creation-based museum concept "combines really bad science and really weak faith."

"Faith and science support each other very well," Schweitzer concluded, "if you let God be God and science be science."

More than a decade removed from that comment, Robert is still irked by such claims.

"God and science," he says, "are not two separate concepts."

Ellie—whose most pressing existential concern involves counting clownfish—leaves Robert and me to tackle the larger questions.

Including my last of the day.

"What do you want museumgoers to take from this place?"

"Well," Robert says, adjusting his glasses while peering up at the *T. rex* just ahead of us, "we do believe that Jesus came to this earth, died on the cross, and rose again for people to have life. And that's ultimately what we want people to understand. Now," he continues, "we don't have a spot here where we make people kneel or parade them over the Bible. We just want them to consider this perspective because we think it's important."

Dinosaurs, he continues, are an essential part of the larger conversation on Christianity.

"A lot of people who were Christians ended up walking away [from the faith] because they were like, 'Well, how does this all fit?'" he says, waving a hand toward the dinosaur fossils.

If Robert can help them find an acceptable answer to that question, more wayward Christians might return to the fold.

As our conversation winds down, we turn our topics to lighter fare.

"How are we doing on those clownfish?" Robert asks Ellie.

"Eighteen!" Ellie says.

"Hmm . . ." Robert says. "I think that might be a little high. Let's see if maybe some different fish got caught up in your count."

I watch Robert and Ellie gamely backtrack through the museum, re-counting clownfish.

This is one question we can get to the bottom of.

*

Upon checking into our hotel later that afternoon, the front desk attendant asks what brings us to Glendive.

"Dinosaurs," I say.

"Dinosaurs, huh?" says the attendant, leaning in close. "Well, if you want to see dinosaurs, you ought to talk to my buddy Maxx."

So, on the advice of the first stranger, an hour later, Ellie and I park the van outside the house belonging to a second stranger—some guy named Maxx.

As we approach Maxx's chain link fence, we are greeted by perhaps half a dozen yipping dogs.

"Quiet! Hush up now," shouts the man I take for Maxx. The dogs more or less comply, then begin racing figure eights around our legs like Shriners' in their cars.

Maxx, seventy-five, wears jeans and a denim shirt. A member of the Sioux tribe with Assiniboine and Oglala roots, he stands a few inches shorter than me, is soft-spoken, and reserved (except when his dogs need disciplining).

Entering Maxx's living room, we are transported to a larger-than-life cabinet of curiosities. Almost every square inch of the place is home to some artifact: dinosaur bones, sure, but also arrowheads, hand-woven snowshoes, beaded moccasins, a life-size cardboard cutout of John Wayne, and—disconcerting as it is—what appears to be a human hand in a jar.

FIG. 28. Ellie and Maxx Martel explore Maxx's personal collection of artifacts and memorabilia in his home in Glendive, Montana. Author photo.

"I've got some dinosaur stuff over here," Maxx says casually, pointing across the room. "*T. rex* claw, *T. rex* teeth, those are all *Hadrosaur*. I've got a dinosaur egg back there, too, and some *Edmontosaurus*." Ellie and I struggle to take it all in, not only because there is so much but because we're experiencing it within a private home. Maxx's house is devoid of "Do Not Touch" signs. In fact, Maxx insists that we hold as much history as we can.

". . . ammonites, turtle shells," Maxx continues, handing them over. "Oh, and I've got crocodile teeth over here . . ."

But the ancient fossils soon give way to more recent, human-made artifacts.

Maxx points us to his John Dillinger wall, complete with a 1928 Thompson submachine gun.

Then onto a 1907 photograph of Buffalo Bill Cody, a 150-year-old golden eagle feathered headdress handed down from an uncle, and finally a framed letter from the U.S. Marshals thanking Maxx for his assistance in recovering the body of a marshal who'd gone missing in 1879.

"What's the story here?" I ask, pausing at the letter.

"Oh, I was out metal detecting in the river, and I found his badge with a bullet hole through it. Then his wallet. Then him. I took part of his jaw to the police station, and they got the U.S. Marshals involved. His name was Driscoll. He was chasing a horse thief near Fort Custer when he was killed."

"And you . . . found him?" I ask.

"Yup," Maxx says. "The marshals sent me that letter and a patch to say thanks."

Though the story defies believability, the framed letter confirms it is true. And it's just one of the many incredible stories Maxx has in store for us.

"These eagle bones are from when I did my sun dance," Maxx says, nodding to several sharpened bones in a nearby case. "They stick them into your back, and you drag a couple of buffalo skulls behind you."

Ellie's eyes widen.

"Like . . . into *your* back?" she asks.

"Yup. In the skin here," Maxx says, pointing to his back before continuing the tour.

"These arrowheads are from a lost battlefield I discovered right outside Glendive," he says, nodding to some points on the wall. "I contacted historians, including Doug Scott, who did the archaeological stuff for Little Big Horn."

He pauses.

"You know, my great-grandmother was at Little Big Horn," Maxx continues, "or the Battle of Greasy Grass, depending on what you call it. She was ten years old. She helped strip the bodies. And she only knew three words of English."

I take the bait. "What were they?"

"'I kill you,'" Maxx reports.

Further down the wall hangs a black-and-white photograph of Maxx's grandfather, a gunfighter who lived to be ninety-nine. "He rode with Pancho Villa," Maxx says.

Then, onto the raptor claw, the Civil War canteen, the powder horn, and hundred-year-old shoes pulled from a creek. He points to a beartrap he found in the Yellowstone River, which remained unsprung for nearly a century. He shows us shark teeth, a ring of lost keys, a shrunken head, a tomahawk, and rubber hand puppets from a Pizza Hut promotion for *The Land Before Time*.

"Should we talk about the hand in the jar?" I ask, winding our way toward the display case.

"Oh, that?" he says. "That was in an old hospital that closed in 1910. I found it while metal detecting."

Maxx walks us toward an original glass photo of President Teddy Roosevelt, then onto a button-maker from 1886. He directs us toward some bullets from the Battle of Greasy Grass. Then a Bowie knife. And finally a war club, which he places in my hands.

"This club," Maxx says matter-of-factly, "was used to kill eleven women and girls."

"Oh," I say, quickly handing it back to him.

As macabre as much of this history is, for Maxx, it's just history. Not good or bad, or right or wrong; just history. From the Mesozoic era right into the present.

"Why does it matter?" I ask, peering at the vast array of artifacts and memorabilia displayed throughout Maxx's house—dinosaur bones in one corner, bullets and guns in another. "Why hold on to all this history, Maxx?"

"People are trying to change history," Maxx says bluntly. "They're trying to get rid of it."

His collection, he explains, ensures that they can't.

*

Back at the hotel, Ellie and I line up our rows of postcards addressed to friends and family across a couple of tables in the lobby.

"Greetings from Glendive," each postcard begins. "We've had quite a day . . ."

Which is, indeed, an understatement.

By every measure, our day had taken a couple of turns. First, some impromptu yoga with James, followed by theological debates with Robert, and finally a tour of Maxx's private museum.

In all, we ventured a long way from dinosaurs, though we managed to work in a few.

There is more to say than could ever fit on a postcard; still, we try.

I handle most of the writing while Ellie scribbles her name. However, midway through the stack, she becomes restless.

"Dad?" Ellie asks.

"Yeah?"

"Can I write one to Gloria?"

I pause, lifting my eyes and my pen. It's the first time she's mentioned a friend in days. Gloria isn't just any friend; she's one of her oldest and best. Through thick and thin, for better or worse, they've long transformed each other's tears into laughter.

I set Ellie up with a fresh postcard and let her get to work.

"Dear Gloria," she writes. "This is Ellie. Greetings from Glendive. We've had quite a day . . ."

FIG. 29. Following their first failed attempt, B.J. and Ellie finally see the World's Largest Buffalo Monument in Jamestown, North Dakota. Author photo.

# 12
# Take Me Home, Country Roads

**JULY 11–12**

*Glendive,* MT → *Ekalaka,* MT → *Bismarck,* ND → *Eau Claire,* WI

Moments before leaving Glendive for Ekalaka's Carter County Museum—our last stop on the Montana Dinosaur Trail—I receive a message from Brendan Heidner, a reporter for the Glendive newspaper.

"I just came across your social media post about you and your daughter's couple of days in Glendive," his message read. "If you're still in town, I would love to catch you before you head out!"

I check my watch. If we're to make the museum at our appointed time, we have about fifteen minutes to spare—all of which we'll spend eating breakfast in the hotel lobby.

I message him about our rather tight timeline and then, on a whim, give him our hotel's location in case he happens to be nearby.

"I'll be there in five minutes," he replies.

Five minutes later, a bearded, bespectacled twenty-four-year-old Brendan sits across the table from us.

"So," he begins, clicking record on his phone's voice memo app. "Tell me about your trip. What inspired it?"

After two weeks of conducting the interviews, this is the first time we've found ourselves on the other side of the recorder.

"Well . . ." I begin. "Well, I guess . . ."

What should have been an easy question isn't.

The trip had been too many things—more than a quote could bear.

As for what had inspired it, that question proved equally complex.

The most honest answer is that many things had—the prospect of adventure, travel, and dinosaurs, sure, but most of all, unencumbered time on the open road with my daughter. The chance to steer us clear of the daily routine to fulfill my own dinosaur dream—meeting incredible people in incredible places.

I wanted to roam from one small town to another, collect all fourteen MDT passport stamps, sleep in the shadows of Makoshika State Park, and awaken in the morning to a life that felt larger than the one we'd left behind.

I wanted Ellie and me to experience the world we're losing before it's gone. And I wanted to inspire us both to try a little harder to save it.

Days before, while seated poolside at the Great Falls KOA, I'd paged through Loren Eiseley's *The Firmament of Time* and stumbled upon the following passage:

"Since the first human eye saw a leaf in Devonian sandstone and a puzzled finger reached to touch it, sadness has lain over the heart of man," Eiseley wrote. "By this tenuous thread of living protoplasm, stretching backward into time, we are linked forever to lost beaches whose sands have long since hardened into stone. The stars that caught our blind amphibian stare have shifted far or vanished in their courses, but still that naked, glistening thread winds onward. No one knows the secret of its beginning or its end. Its forms are phantoms. The thread alone is real; the thread is life."

What really inspired our trip? I suppose it was the search for the thread, a thread that leads us forward into our future and backward toward a past with so much to teach.

But since that, too, is more than a quote can bear, I give Brendan a simpler answer.

"Well," I say, "I guess I just wanted to spend some time with my daughter."

*

One hundred and eight miles outside Glendive, the road rises and falls toward Ekalaka. Situated in Montana's southeastern corner, Ekalaka (population 399) is home to the Carter County Museum, a few schools, a cemetery, and several churches. But the museum is what brings us to town—not only to receive our final passport stamp but also due to its reputation as one of the highlights along the MDT.

Established in 1936 by the Carter County Geological Society, for decades, the museum was housed in the basement of the local high school, only moving to its current location in 1980. Like Fort Peck, Ekalaka received its fifteen minutes of fame in the pages of *LIFE Magazine*. A 1954 article describes the town as a fossil hunter's paradise: "Fossil-hunting, usually a specialty for experts, is the hobby of a whole community in Ekalaka," the reporter wrote.

The article featured Marshall Lambert, the town's amateur paleontologist, high school science teacher, and director of the Carter County Museum, which he managed from 1946 to 1996. A lifelong learner, Lambert obtained several degrees—though none in geology or a paleontology-adjacent field. On the advice of relatives, he majored in business ("geologists were starving to death," he'd been told) and, after a combat tour as a pilot during World War II, received his teaching certificate and a master of science degree. He began teaching at Carter County High School in 1946 (the same school he'd graduated from in 1932), retiring from the role in 1975. All the while, he dug up dinosaurs, enrolled in additional science classes, and studied alongside fossil preparators to ensure he could foreground science with every fossil find.

Lambert is the ideal example of what a passionate amateur might provide the paleontology community: eyes on the ground, local knowledge, and a commitment to partnership. He didn't dedicate his life to dinosaur fossils for fame or fortune. He did it because he liked dinosaurs and loved and his community and saw a way to better both by combining them.

In 1996, at eighty-one, Lambert received a bit of fame nonetheless, earning the Paleontological Society's Harrell L. Strimple Award, recognizing outstanding achievement in paleontology by an amateur. The society commended Lambert for "collecting, preserving, displaying, and sharing of specimens, his long record of friendly assistance to generations of student and professional paleontologists in field and laboratory work, and his advancement of paleontological education of the public through personally building and maintaining a regional museum."

With Lambert's legacy in mind, we enter the fruits of his labor, the Carter County Museum. Ellie and I are greeted by thirty-year-old gift shop manager Pat Rouane, who towers over everything but the dinosaur displays. He wears a trucker hat that doesn't come close to containing his full head of hair.

"You're the ones writing the book," Pat says. "Well, welcome! Let me show you guys around."

Pat guides us into the Lambert Dinosaur Hall (named after Marshall Lambert, of course), where we soon come face to snout with a fully articulated *Edmontosaurus* discovered thirty miles west of town in the summer of 1938. As museum director Sabre Moore writes, unearthing the *Edmontosaurus* was nothing short of a "community affair." For two months, locals lugged shovels and refreshments to the dig site, where they enjoyed long, languishing afternoons engaged in an activity that fit somewhere between a scientific expedition and a town potluck. By summer's end, they loaded their fossils onto a trailer and drove them back to town, where one of the world's largest publicly displayed *Edmontosauruses* found its new home in the Carter County High School basement.

While the Carter County Museum is best known for its *Edmontosaurus* (the signature dinosaur for their passport stamp), it has also recently garnered attention for its Dino Shindig—a weekend-long dinosaur-filled extravaganza featuring dig site tours, lectures from paleontologists, kids' activities, a barbecue, and a street dance. Founded in 2013, this event brings together dinosaur lovers from across the region and serves as a financial boon to the town. In 2019 the event attracted more than nine hundred visitors, contributing over $18,000 to the local economy.

The Dino Shindig also embodies Marshall Lambert's inclusive spirit. Everyone is welcome, and most years, nearly everyone in the state's paleontology community shows up.

Within minutes of our tour, it's clear that Pat's knowledge extends well beyond managing the gift shop. His expertise transcends eras; he can as easily talk about a *Smilodon* skull as the grooves in an ancient horse tooth.

When I ask Pat if he, like so many others, is also part of the *Jurassic Park* generation, he chuckles.

"From what I'm told, I've always loved dinosaurs," he says. "I remember getting *The Land Before Time* on cassette one Christmas. And I had this hard toy *Brachiosaurus*—which was still a dinosaur back then—and it had this battery pack on the bottom. So you'd throw in like twelve AA batteries, and it would run for like two minutes," he laughs. "I'd sit there watching the movie and playing with that toy. It was great."

Later, when he watched *Jurassic Park*, he was entranced by the "real-life" dinosaurs that moved not by batteries but by the magic of cinema.

"You know Jack Horner helped with that movie," a relative once shared with him.

"Who?" Pat asked.

"A paleontologist up in Bozeman," the relative explained. "You can study dinosaurs there too one day."

Years later, he did.

After graduating in 2017, Pat tried an array of jobs. He worked in a bakery, a hardware store, and in a ranger-like role in the foothills of the Rocky Mountains. In 2021 he was able to put his paleontology degree to better use when he was hired at the Carter County Museum. Though Pat's not getting rich (few in the paleontology community ever do), he loves the work, which allows him to study ancient flora and fauna alongside people as committed as he is.

In an era in which bank accounts and social media followers seem to be our primary currencies, Pat's professional life, in part, is guided by wonder. Throughout our trip, we've met many others who live by wonder too. Folks like Tom Hebert, Zach Perry, Dave Trexler, and Dinodave Fuqua. But also Kerry Libby, the many museum directors, and even Jack Horner. Some of these folks are professional paleontologists, others are amateurs, and others wouldn't even call themselves that. Regardless of their designations, they all share a deep reverence for all things of the Mesozoic era and a deep commitment to fully understanding what that era might teach us about today.

But dinosaur admirers aren't the only passionate folks we've met on this trip. I think of Richard and Diane of the Rivertown Rounders, who sing and strum at the Great Falls campground for months, stirring travelers' souls.

And of Bynum teacher Susan Luinstra, who taught and danced for decades to inspire her students. And of Maxx Martel in Glendive, who preserves the history others might forget.

Almost everyone we met fostered community in the places they call home. Like all those folks in Roundup willing to loan Tom Hebert a truck or a tire. And the people of Rudyard, who welcomed us to town with Klondike Bars. The woman at the front desk at the Lewis and Clark Interpretive Center, who enlisted Ellie as a Junior Ranger. The oil change technician in Great Falls, who saw a faulty windshield wiper and fixed it.

Modest as they were, such kindnesses seem like an evolutionary step in the right direction, particularly compared to the cruelties of the late nineteenth-century Bone Wars. What a tragedy that the scientific contributions of Edward Drinker Cope and Othniel Charles Marsh have become overshadowed by their feud, that their legacy now lives on as a cautionary tale.

Today, we can choose a different path.

Staring into the eye sockets of the *Edmontosaurus*, I see a bit of myself reflected—not in a species sense, but in terms of mortality. Specifically, I see how a creature that once "ruled the world" is now reduced to fossils. A similar fate awaits us all.

While fossils have taught us plenty about dinosaurs, we're still discovering what dinosaurs might teach us about ourselves.

Perhaps it's less about their extinction than its aftermath. Sixty-six million years ago, an asteroid struck and killed the bulk of the world's animals, but what about those who survived?

Might their survival provide clues to our own?

Today, we are witnessing something that has never been seen by human eyes—the beginning of our end. If you ask scientists, they will tell you that the sixth mass extinction has already begun. While we bear the bulk of the blame for our current circumstances, remember that we humans played no role in the five extinctions that came before. As devastating as extinction is, it is also a natural event.

But make no mistake, it's different this time. For the first time in over 4.5 billion years of planetary history, a species' actions are directly linked to that species' demise.

And yet, lamenting our actions will not reverse them.

I find hope in knowing that humans love a good comeback story.

But do we love them enough to buy ourselves a little more time? Can our passion for paleontology provide us with much-needed answers? Can science lead us toward extending our survival?

"Why does it matter?" I ask Pat in the shadow of the *Edmontosaurus*. "Why dedicate your life to dinosaurs?"

"Why dedicate your life to being an accountant?" Pat chuckles. "Why does that matter?" He pauses, giving the question some real thought.

"Look, humans are always looking into the future," he says, "and there's a lot of science that looks toward the future too: engineering, chemistry, even history. But we're always digging up the past. And [that past] is reflective of how we view ourselves in the future."

*How do we view ourselves in the future?* I wonder.

In those moments when we're bullish enough to envision our future at all.

*

Back in the gift shop, the woman working the register reaches for our Montana Dinosaur Passport. We hand it over for the last time. She flips expertly through the pages, positions her *Edmontosaurus* stamp to the final space, and presses down.

If I had expected balloons to drop from the ceiling, I would've been deeply disappointed. What we receive instead is a round of high-fives from all those present. It is enough. It is more than enough.

In the time it takes for a stamp to dry, we have done what we set out to do. The only place left to go is home, which we'll return to within twenty-four hours.

"You know," I say, buckling back into the van, "few have accomplished what we just did."

"Really?" Ellie asks, paging through the passport to admire the stamps.

"Well, maybe a couple thousand a year," I admit. "But how does it feel to be one of them?"

"Good," she says. "Really good. Plus, no more museums for a while."

I shoot her a look in the rearview.

"What?" she laughs. "There were a lot of museums."

"And a lot of dinosaurs," I remind.

"Yeah, yeah," she concedes, taking one last glimpse of Ekalaka, "a lot of dinosaurs too."

I glance in the rearview to catch her fitting her headphones to her ears, then listen as she hums along to a song.

Fresh from our victory of a full passport, I realize how lucky I am. Despite the perils of our planet, I couldn't have picked a better copilot to join me wherever our journey takes us next.

Which is probably what some daddy dinosaur thought about his daughter as an asteroid lit up the sky.

*

Two hours outside of Ekalaka, we arrive at Painted Canyon Overlook in Theodore Roosevelt National Park—and try not to get blown away. The winds have picked up mercilessly, whipping the visitor's center flag far from its flagpole. Ahead of us, badland formations stretch on for miles like a miniature Grand Canyon.

The view is breathtaking and monotonous—a description fitting for nearly every mile we've logged throughout this trip. Even on the longest and least scenic stretches of the Great Plains, there was always something to admire out the window. Something that nourished our souls and fueled our curiosity. The trick is to see every inch of the country on its own terms. Not to compare the changing landscape against what came before or what might come next, but to savor the present view.

The array of license plates in the overlook parking lot confirms that we're hardly the only ones enjoying the view. However, I'm particularly interested in the dozen or so bicyclists, each of whom stoically clip-clop between the bathrooms and the visitor center, using any structure they can find for a windbreak.

It's bad enough driving through twenty-mile-per-hour winds; I can't imagine trying to pedal through them. I strike up a conversation with the cyclist closest to us, an unshaven man in his early sixties.

"Where are you all headed?" I ask.

"Seattle," the man says.

"Washington?" I ask incredulously.

"We're raising awareness for cancer," he explains. "We started in Minneapolis, and we're headed to Seattle."

"*Washington?*" I repeat.

"That's the plan," he says.

"I'm sorry . . . you're riding to Seattle, Washington? On bikes?"

He smiles stoically.

"How about you two?" he asks. "Where are you headed?"

"Home," Ellie says. "We just did the Montana Dinosaur Trail!"

"We're raising awareness about . . . dinosaurs," I explain.

"All right then," the man chuckles, snapping his helmet back to his head. "Well, safe travels."

"You too," I say. "Watch out for the . . . wind."

He gives us a wave.

Back on the highway, I say, "You know, that might make for a fun adventure someday."

"What would?" Ellie asks.

"You and me, biking across the country . . ."

Ellie rolls her eyes.

But then, after a moment's thought, she surprises me by saying maybe. Someday. Just not someday soon.

"We've got nothing but time," I assure her. Hoping that it's true.

*

Back in Bismarck, we return to the same KOA where it all began two weeks before. Nothing has changed except the temperature, which has dropped since our last visit. My weather app informs me that the night calls for low fifties and rain.

I'm not thrilled by the prospect of another chilly night in the tent. Yet when we check in at the front desk, I'm surprised to learn that six months before—amid all my hastily made travel arrangements—I'd had the foresight to book a cozy cabin, nearly identical to the one in Great Falls.

"Here ya go," says the woman behind the front desk, handing us the key. "Your cabin's the third one on the left."

Ellie and I stare at the key as if having been handed a *Pachycephalosaurus* skull.

It is not the presidential suite, though we tremble with gratitude upon entering. There is the heater, the mini fridge, and the desk. We are in Great Falls all over again, minus the Rivertown Rounders.

That night, Ellie hops in the pool—temperatures be damned. Then we roam a nearby grocery store to buy supplies for lunch in the van for the following day. We purchase a root beer too, and a Charleston Chew, then top off the tank at the same gas station where we topped it off before leaving for Fort Yates.

One last loop of the campground. One last walk to the bathrooms to brush our teeth.

On the way back to our cabin, we take a brief detour to pay respects to our former campsite.

"We sat right there," Ellie says, nodding to the bench, "and we read *Ramona the Brave*."

"I'm glad Ramona was brave," I say, "because I sure wasn't feeling too brave back then."

"Why not?"

"I don't know. I was nervous, I guess."

"About what?"

"All kinds of things. What if we got sick? Or what if we had car trouble? Or what if we lost our Montana Dinosaur Trail passport?"

"Oh," Ellie says. "That would have been bad. Losing our passport, I mean."

"Right?" I say. "Then we'd have to go back and do it all over again!"

She laughs. Her father is out of his mind.

So many things could have gone wrong but didn't. We didn't get lost, caught in a storm, or bitten by a snake. Perhaps the worst thing we endured was a broken jar of strawberry jam, and even then, we were spared the bears.

While I learned plenty over the past two weeks, the most important lesson was this:

We can't live our lives fearing worst-case scenarios. All we can do is live our lives.

Overhead, billion-year-old stars shine upon us. As they did for the dinosaurs too.

*

We wake a few minutes past 5:00 a.m., anxious to hit the road. The rain has stopped, returning the world to a blackened, predawn slumber. Ellie and I sleepwalk to the bathroom, punching the codes into our separate doors. I splash water on my face, then peer into the mirror to see the bedraggled version of me peering back. The bedraggled, happy version.

Outside the bathroom, I wait for Ellie to emerge. I miss everyone—my wife, son, daughter, and the dog. But I try to savor our final dawn in this campground in North Dakota, where, odds are, we'll never be again.

The door to the women's bathroom opens, revealing Ellie in silhouette. She seems taller than the last time we were here. Maybe she is.

"Ready?" she asks, shaking sink shower water from her hair.

I smile and nod. What choice do I really have?

*

Shortly after 6:00 a.m., I see the billboard announcing the World's Largest Buffalo Monument in Jamestown, North Dakota.

Due to my inability to properly follow roadside attraction signage, we'd mostly missed it on the first day of our trip.

We sure as hell aren't about to miss it again.

"Pit stop," I say, taking the exit.

"Why?" Ellie asks.

"You'll see."

The World's Largest Buffalo Monument takes on a new level of majesty when not viewed from behind a fence and obscured by bushes and trees.

At twenty-six feet tall, it's nearly as big as a *Brachiosaurus*. It stands stately atop its concrete slab, the sky stretching in all directions. A puddle has formed beneath its beard, the result of the previous day's rain. Ellie stands in the puddle and tiptoes toward the beard, barely reaching it.

"We did it!" she shouts with an enthusiasm that rivals the previous day's trail completion. "We saw the World's Largest Buffalo!"

"The World's Largest Buffalo *Monument*," I clarify.

We position ourselves before it as I fumble with the camera setting on my phone.

"Okay, smile and say, 'We love dinosaurs!'"

"Do I have to?" she whispers.

"Ellie!"

"I'm kidding," she laughs. "I actually do love dinosaurs! Mostly."

A few hours from home, as Ellie and I listen to our last *Boxcar Children* book of the trip, Mr. Alden—the children's grandfather—says to one of the boys, "Try to look forward to everything, even ends."

We don't have to try very hard. After thirteen days and 2,500 miles, all we want in the world is to reunite with the people we love.

"Dad?" Ellie says somewhere in Minnesota.

"Yeah, hon?"

"Can you put on 'Take Me Home, Country Roads'?"

"You know," I say, snapping my fingers, "now that sounds like a great title for a country and western song!"

"Da-ad!" she moans.

"Fine, fine," I laugh. "One more time from the top."

*

As we pull into the driveway later that afternoon, I notice that our chalk outlines have worn away. The hose or the rain has erased all traces of us. Thank goodness we got back when we did.

I've barely put the van in park before Ellie leaps out the door, racing into the house to hug every human within arm's reach.

Henry and Millie don't quite know what to make of their sister, who lovingly dogpiles them to the kitchen floor in a mound of hugs. Dutifully, she performs her service as the middle child, holding them all together.

"How was it?" Henry asks, removing himself from the pile.

"It was great!"

"What was great?" asks three-year-old Millie, still trapped in the dogpile. "Wait . . . where you been, Ellie?"

Laughing, Ellie stands and turns to see Meredith, her arms outstretched.

"Mom!" Ellie calls, falling into them.

"We missed you so much," she says.

"We missed *you* so much," Ellie echoes.

Barking, the dog heralds our return, then races around the living room as the children give chase.

"Welcome home," Meredith says, kissing me. I, too, collapse into her arms.

"Remind me never to leave home again," I say, exhausted.

"Like that'll work," she laughs.

As the children and dog make their third lap around the living room, I'm reminded of everything worth fighting for—and reminded, too, that one form of legacy is embedded in the people we love.

In the end, we are all reduced to the fossil record.

What matters most is all that comes before.

# Epilogue

In the fall of 2023, my former student Tom Hebert learned that the Montana Dinosaur Trail had denied Roundup's Musselshell Valley Historical Museum's bid to join the MDT. MDT members were concerned about Tom's affiliation with the bid. If the museum reapplies without Tom and Earth Sciences Foundation attached, I suspect they'll have a more favorable outcome.

Tom had hoped to help secure the Musselshell Valley Historical Museum a spot on the trail, yet ironically, his "help" may have doomed them. After weeks of mulling his options, Tom found a different way to help the people of Roundup—by severing his formal agreement with the museum.

"While it has been [Earth Sciences Foundation's] hope to be a part of [Musselshell Valley Historical Museum's] entrance into the Dinosaur Trail," Tom wrote the museum, "with the given concerns expressed at your board meeting on November 12, 2023, about ESF, and the feeling that MVHM will have a better chance gaining entrance to the Dinosaur Trail as an independent entity, we are terminating our formal agreement."

While some locals questioned Tom's intentions regarding locally found fossils (namely, that he might abscond with them in the night), Tom clarified in writing that the museum should "retain the locally found fossils of their choosing and display them as deemed necessary." He added that Earth Sciences Foundation would "take reasonable steps to further assist MVHM's future Dinosaur Trail efforts if requested," though so far, no one has asked for Tom's help.

Later, when I asked Tom if it was difficult to distance himself and his foundation from Roundup, he paused before answering.

"Not really," he said at last. "If I was serious about trying to help get Roundup on the Montana Dinosaur Trail, then leaving was the right thing to do."

Over the past several months, Tom's turned his attention beyond Roundup and the MDT, focusing instead on developing new ways to track fossils using radio-frequency identification tags (RFID). Imagine a tiny computer chip attached to a bone. At the click of a button, the tags allow anyone anywhere to access every data point from each tagged fossil.

"The technology exists," Tom says, walking me through his data. "The hard part will be convincing people to play nice and share it."

For the past several months, he's been widely sharing his RFID fossil-tagging technology methods at conferences and museums and, most recently, with a representative from the Smithsonian Natural Museum of Natural History.

A month or so after we camped with Tom near Big Wall, he officially re-enrolled at our university. He assured me that the third time was the charm, and this time, he's not leaving without a degree.

We regularly passed each other in the hallways throughout the fall semester, sharing smiles, waves, and rushed conversations as we raced to our various classes. I hoped he was succeeding in the classroom, but I didn't know for sure.

Until January of 2024, when he forwarded me an email from the university confirming that he'd made the Dean's List.

For the first time in his life.

# Notes on the Sources

This book was written from various sources, including firsthand accounts, scholarly research, and online and newspaper articles. What follows is a list of the sources I most heavily relied upon while crafting individual chapters. The sources are ordered according to the information's placement within each chapter. If a source was employed multiple times throughout a chapter, I listed it only upon its initial use.


**PROLOGUE**

Preston, "The Day the Dinosaurs Died."

Klotz, "Asteroid May Have Killed Dinosaurs Quicker than Scientists Thought."

"Introduction to Human Evolution."

Parson, "A Moment That Changed Earth."

Renne et al., "Time Scales of the Critical Events around the Cretaceous-Paleogene Boundary."

Kaiho, "Global Climate Change Driven by Soot at the K-Pg Boundary as the Cause of Mass Extinction."

Greshko, "These Early Humans Lived 300,000 Years Ago—But Had Modern Faces."

Johnson, "That Thou Canst Not Stir a Flower / Without Troubling of a Star."

"What Does It Mean To Be Human?"

The Laurie Berkner Band, "We Are the Dinosaurs."

Schweitzer, "Dinosaurs Reveal Clues About Adaptation to Climate."

"How Do We Know Climate Change Is Real?"

Daigle and Janicki, "Extinction Crisis Puts 1 Million Species on the Brink."

Spears, "The World's Population May Peak in Your Lifetime."

The Montana Dinosaur Trail.

**1. A BONE TO PICK**

Wallace, *The Bonehunters' Revenge*.

Storrs, "*Elasmosaurus platyurus* and a Page from the Cope-Marsh War."

Osborn, *Cope, Master Naturalist.*
Schuchert, *Biographical Memoir of Othniel Charles Marsh.*
Henriques, "The Bitter Dinosaur Feud at the Heart of Paleontology."

## 2. EMBRACE THE EVOLUTION

Thomas Hebert, personal interview, June 30, 2023, July 1–4, 2023.
Thomas Hebert, personal correspondence, January 28, 2021.
North Dakota Native Tourism Alliance, "Standing Rock Sioux Tribe."
Kerry Libby, personal interview, June 30, 2023.
Standing Rock Sioux Tribe Council, "Standing Rock Sioux Code of Justice Paleontology Resource Code."
Guy, "*T. rex* Skeleton Sells for $31.8 Million Setting New World Record."
Browne, "Tyrannosaurus Skeleton Is Sold to a Museum for $8.36 Million."
Mayor, *Fossil Legends of the First Americans.*
"*Dryptosaurus.*"
Bakker, "Dinosaur Renaissance."
Balter, "Which Came First."

## 3. DOLLARS FOR DINOS

"Big Mike."
Victor Bjornberg, personal interview, June 6, 2023.
Data provided by Victor Bjornberg.
"Montana Population, 1900–2023."
"Montana Paleontology Top Ten."
Montana Office of Tourism, "2021 Economic Impact."
Zach Perry, personal interview, July 1, 2023.
University of Notre Dame, "Saving Peck's Rex."
"How *T. rex* 'Lost' Its Third Claw."
Associated Press, "Jurassic Park Paleontologist Retiring from Museum He Built."
Society of Vertebrate Paleontology, "SVP Ethics Code."
Clayton Phipps, personal interview, June 7, 2023.
Pantuso, "Perhaps the Best Dinosaur Fossil Ever Discovered."
Kaplan, "Dueling Dinosaurs Fizzle in Auction."
Cornwall, "Court Rules 'Dueling Dinos' Belong to Landowners, in a Win for Science."
Hanson, "Montana Legislature Clarifies Ownership Rights of Fossils."
Nielson, "The 'Dueling Dinosaurs' Fossil Shows a *T. rex* and *Triceratops* in a Possible Fight."
Hodges, "Dinosaur Wars."

Josey Montana, personal interview, July 1, 2023.

Tim Busby, personal interview, July 1, 2023.

### 4. TIRES MAKE GOOD NEIGHBORS

Thomas Hebert, personal interview, June 30, 2023, July 1–4, 2023.

Ambrose, "35 Quotes About Montana to Transport You to the Big Sky."

Beth Hebert, personal interview, July 2–3, 2023.

"What Is a Paleontologist?"

Beye, "Our Story: Roundup, Montana."

### 5. WHERE THE DEER AND THE DINOSAURS PLAY

Sternberg, *The Life of a Fossil Hunter.*

Beth Hebert, personal interview, July 4, 2023.

Thomas Hebert, personal interview, June 30, 2023, July 1–4, 2023.

### 6. OF MAMMOTHS AND MEN

Center for Digital Research in the Humanities, "Journals of the Lewis and Clark Expedition."

Edwards, "Thomas Jefferson's Secret Reason for Sending Lewis and Clark West."

Osterloff, "Dinosauria."

"Missouri River."

Viola, "Native Peoples, Lewis and Clark, and Mapmaking."

"Rainbow Falls (Missouri River)."

Randall, *The Monster's Bones.*

Moore, "Rural Museums."

Richard Baker, personal interview, July 3–4, 2023.

Diane Stinger, personal interview, July 3–4, 2023.

### 7. MATERNAL INSTINCT

"The Front."

Graetz, "This Is Montana."

"These Small Town Students Sing and Dance Every Morning to Learn How to Get Along."

Susan Luinstra, personal interview, July 6, 2023.

Sean Doyle, personal interview, July 6, 2023.

"Museum of the Rockies."

Schontzler, "Museum of the Rockies Adds $48 Million to Local Economy."

"Bozeman, MT."

"Choteau, MT."

Dave Trexler, personal interview, July 6, 2023.

Trexler, "Marion Brandvold Celebrates 100th Birthday."

Horner and Gorman, *Digging Dinosaurs.*

Horner and Makela, "Nest of Juveniles Provides Evidence of Family Structure Among Dinosaurs."

### 8. HIT THE ROAD FOR JACK

Horner and Gorman, *Digging Dinosaurs.*

"Jack Horner, Paleontologist."

Horner, "An Intellectual Autobiography."

"Chief Joseph Surrenders."

Samantha French, personal interview, July 7, 2023.

Lela Patera, personal interview, July 7, 2023.

Marin, *James J. Hill and the Opening of the Northwest.*

Lila Redding, personal interview, July 7, 2023.

Jack Horner, personal interview, July 7, 2023.

### 9. LEONARDO AND *MATILDA*

Sternberg, *The Life of a Fossil Hunter.*

Dave Trexler, personal interview, October 19, 2023.

Associated Press, "Malta Welcomes Dinosaur's Return."

"Fort Peck, Montana."

*Fort Peck Dam.*

U.S. Army Corps of Engineers, "Historical Vignette."

"The Jewel of the Prairie."

### 10. AMATEUR HOUR

Hendrix, "The Nation's *T. rex.*"

Sachar, *There's a Boy in the Girls' Bathroom.*

Sue Dalbey, personal interview, July 9, 2023.

Kreier, "How Many *T. rex* Ever Existed?"

"David H. Koch Hall of Fossils—Deep Time."

"Welcome to Dinosaur Country!"

Jacque Gregg, personal interview, July 9, 2023.

"Jordan, Montana."

"Barnum Brown."

Randall, "He Was An All-Time Genius at Finding *Tyrannosaurus rexes.*"

Dingus and Norell, *Barnum Brown.*
"Barnum Brown Dies at 89."
Osterloff, "The World's First Dinosaur Park."
Coules and Benton, "The Curious Case of Central Park's Dinosaurs."
"Mining for Mammoths in the Bad Lands."
Visit Rapid City Staff, "How a Depression-Era Park Became a Favorite Rapid City Attraction."
"Dinoland Pavilion at the World's Fair."
Dinodave Paleo Adventures, "Secret Dinosaur Bones Along the Diane Gabriel Trail."
Dave Fuqua, personal interview, July 9, 2023.
Society of Vertebrate Paleontology, "SVP Ethics Code."

### 11. OF FAITH AND FOSSILS

James Kelley, personal interview, July 10, 2023.
Healy, "Paleontologists Take Issue with Creation-Based Museum."
Brenan, "40% of Americans Believe in Creationism."
Robert Canen, personal correspondence, June 23, 2023.
Robert Canen, personal interview, July 10, 2023.
Peppe and Deino, "Dating Rocks and Fossils Using Geologic Methods."
Senter, "Radiocarbon in Dinosaur Fossils."
Water Science School, "The Atmosphere."
Cho et al., "The Tiger Genome and Comparative Analysis with Lion and Snow Leopard Genomes."
"Meet Our Team."
Funk, "Perception of Conflict Between Science and Religion."
Inbody, "Dinosaurs, the Bible, and a Glendive Museum."
Azarian, "Why Carl Sagan Believed That Science Is A Source of Spirituality."
Maxx Martel, personal interview, July 10, 2023.

### 12. TAKE ME HOME, COUNTRY ROADS

Brendan Heidner, personal correspondence, July 11, 2023.
Brendan Heidner, personal interview, July 11, 2023.
Eiseley, *The Firmament of Time.*
"The Town That Hunts Bones."
Holland and Hartman, "Presentation of the Harrell L. Strimple Award of the Paleontological Society to Marshall E. Lambert."
Pat Rouane, personal interview, July 11, 2023.
Moore, "Rural Museums."

# Bibliography

"Age of Dinosaurs." National Park Service. https://www.nps.gov/subjects/fossils/dinosaurs.

Ambrose, Jen. "35 Quotes About Montana to Transport You to the Big Sky." Montana Discovered, December 17, 2020. https://montanadiscovered.com/quotes-about-montana/.

American Rivers. "Missouri River." https://www.americanrivers.org/river/missouri-river/.

Associated Press. "Jurassic Park Paleontologist Retiring from Museum He Built." *Denver Post*, May 21, 2016. https://www.denverpost.com/2016/05/31/jurassic-park-paleontologist-retiring-from-museum-he-built/.

———. "Malta Welcomes Dinosaur's Return." *Missoulian*, July 7, 2001. https://missoulian.com/malta-welcomes-dinosaur-s-return/article_960ee21d-b364-5daa-b7a7-873293bad86c.html.

Azarian, Bobby. "Why Carl Sagan Believed That Science Is A Source of Spirituality." Big Think, February 9, 2023. https://bigthink.com/thinking/why-carl-sagan-believed-that-science-is-a-source-of-spirituality/.

Bakker, Robert T. "Dinosaur Renaissance." *Scientific American* 232, no. 4 (1975): 58–79. http://www.jstor.org/stable/24949774.

Balter, Michael. "Which Came First: The Dinosaur or the Bird?" *Audubon Magazine*, January–February 2015. https://www.audubon.org/magazine/january-february-2015/which-came-first-dinosaur-or-bird.

"Barnum Brown: The Man Who Discovered Tyrannosaurus Rex." Produced by James Sims and Jill Bauerle. American Museum of Natural History, 2011. https://www.amnh.org/explore/videos/dinosaurs-and-fossils/barnum-brown-the-man-who-discovered-tyrannosaurus-rex.

"Barnum Brown Dies at 89." *New York Times*, February 6, 1963. https://timesmachine.nytimes.com/timesmachine/1963/02/06/89518129.html?pageNumber=9.

Benton, Michael J. *The Dinosaurs Rediscovered: How a Scientific Revolution Is Rewriting History.* Thames and Hudson, 2019.

Beye, Wendy. "Our Story: Roundup, Montana." Visit Roundup. https://www.visitroundup.com/post/our-story-roundup-montana.

"Big Mike." Museum of the Rockies. https://museumoftherockies.org/exhibitions/big-mike.

"Bozeman, MT." Data USA. https://datausa.io/profile/geo/bozeman-mt/.

Bradley, Lawrence. "Dinosaurs and Indians: Fossil Resource Disposition of Sioux Lands, 1846–1875." *American Indian Culture and Research Journal* 38, no. 3 (2014).

Brannen, Peter. *The Ends of the World: Volcanic Apocalypses, Lethal Oceans, and Our Quest to Understand Earth's Past Mass Extinctions.* Ecco, 2017.

Brenan, Megan. "40% of Americans Believe in Creationism." Gallup, July 26, 2019. https://www.scribd.com/document/760299198/1-Megan-Brenan-40-of-Americans-Believe-in-Creationism.

Browne, Malcolm. "Tyrannosaurus Skeleton Is Sold to a Museum for $8.36 Million." *New York Times,* October 5, 1997. https://www.nytimes.com/1997/10/05/nyregion/tyrannosaur-skeleton-is-sold-to-a-museum-for-8.36-million.html.

Brusatte, Stephen. *The Age of Dinosaurs.* Quill Tree, 2021.

———. *The Rise and Fall of the Dinosaurs: A New History of a Lost World.* William Morrow, 2018.

———. *The Rise and Reign of the Mammals: A New History, from the Shadow of the Dinosaurs to Us.* Mariner, 2022.

Center for Digital Research in the Humanities. "Journals of the Lewis and Clark Expedition." https://lewisandclarkjournals.unl.edu/.

Chen, Chaomei. "On the Shoulders of Giants." *Mapping Scientific Frontiers: The Quest for Knowledge Visualization.* Springer, 2003, https://doi.org/10.1007/978-1-4471-0051-5_5.

"Chief Joseph Surrenders." Library of Congress. https://www.loc.gov/item/today-in-history/october-05/.

Cho, Y., L. Hu, H. Hou, et al. "The Tiger Genome and Comparative Analysis with Lion and Snow Leopard Genomes." *Nature Communications* 4, no. 1 (2013): 2433, https://doi.org/10.1038/ncomms3433.

"Choteau, MT." Data USA. https://datausa.io/profile/geo/choteau-mt/.

Cope, E. D. "On Some Extinct Reptiles and Batrachia from the Judith River and Fox Hills Beds of Montana." *Proceedings of the Academy of Natural Sciences of Philadelphia* 28 (1876): 340–59. https://www.biodiversitylibrary.org/item/84760.

Cornwall, Warren. "Court Rules 'Dueling Dinos' Belong to Landowners, in a Win for Science." *Science,* May 22, 2020. https://www.science.org/content/article/court-rules-dueling-dinos-belong-landowners-win-science.

Coules, Victoria, and Michael J. Benton. "The Curious Case of Central Park's Dinosaurs: The Destruction of Benjamin Waterhouse Hawkins' Paleozoic Museum Revisited." *Proceedings of the Geologists' Association* 134, no. 3 (2023): 344–60, https://doi.org/10.1016/j.pgeola.2023.04.004.

Daigle, Katy, and Julia Janicki. "Extinction Crisis Puts 1 Million Species on the Brink." *Reuters*, December 23, 2022. https://www.reuters.com/lifestyle/science/extinction-crisis-puts-1-million-species-brink-2022-12-23/.

"David H. Koch Hall of Fossils—Deep Time." National Museum of Natural History. https://naturalhistory.si.edu/exhibits/david-h-koch-hall-fossils-deep-time.

Dingus, Lowell, and Mark Norell. *Barnum Brown: The Man Who Discovered Tyrannosaurus Rex*. University of California Press, 2010.

Dinodave Paleo Adventures. "Secret Dinosaur Bones Along the Diane Gabriel Trail 'Special' Episode." YouTube, October 23, 2022. https://www.youtube.com/watch?v=c8xi5voozlo.

"Dinoland Pavilion at the World's Fair: Museum Connections." American Museum of Natural History, April 22, 2022. https://www.amnh.org/explore/news-blogs/news-posts/dinoland-pavilion-1964-worlds-fair.

"*Dryptosaurus*." American Museum of Natural History. https://lbry-web-007.amnh.org/digital/index.php/items/show/90713.

Edwards, Phil. "Thomas Jefferson's Secret Reason for Sending Lewis and Clark West: To Find Mastodons." *Vox.com*, April 13, 2015. https://www.vox.com/2015/4/13/8384167/thomas-jefferson-mastodons.

Eiseley, Loren. *The Firmament of Time*. University of Nebraska Press, 1999.

Fiffer, Steve. *Tyrannosaurus Sue: The Extraordinary Saga of the Largest, Most Fought Over T-Rex Ever Found*. W. H. Freeman, 2001.

"Fort Peck, Montana." Wikipedia. https://en.wikipedia.org/wiki/Fort_Peck,_Montana.

*Fort Peck Dam*. Montana PBS, 2012. https://www.pbs.org/show/fort-peck-dam-jsm/.

"Fort Peck Lake Reservoir and Recreation Area." Visit Montana. https://www.visitmt.com/listings/general/lake/fort-peck-lake-reservoir-and-recreation-area.

"The Front." Old Trail Museum. https://www.oldtrailmuseum.org/the-front.

Funk, Cary. "Perception of Conflict Between Science and Religion." Pew Research Center, October 22, 2015. https://www.pewresearch.org/science/2015/10/22/perception-of-conflict-between-science-and-religion/.

Graetz, Rick, and Susie Graetz. "This Is Montana: Choteau Is Scenic Heart of Rocky Mountain Front's Abrupt Grandeur." *Billings Gazette*, June 6, 2023.

Greshko, Michael. "These Early Humans Lived 300,000 Years Ago—But Had Modern Faces." *National Geographic*, June 7, 2017. https://www.nationalgeographic.com/history/article/morocco-early-human-fossils-anthropology-science.

Guy, Jack. "*T. rex* Skeleton Sells for $31.8 Million Setting New World Record." *CNN.com*, October 11, 2020, https://www.cnn.com/style/article/stan-t-rex-skeleton-auction-scli-intl-scn/index.html.

Hanson, Amy Beth. "Montana Legislature Clarifies Ownership Rights of Fossils." *Associated Press*, March 22, 2019. https://apnews.com/general-news-2bbfc0515ca1422ea866d962ee9897e9.

Healy, Donna. "Paleontologists Take Issue with Creation-Based Museum." *Billings Gazette*, October 10, 2009.

Hedeen, Stanley, and John Mack Faragher. *Big Bone Lick: The Cradle of American Paleontology*. University Press of Kentucky, 2015.

Hendrix, Steve. "The Nation's *T. rex*: How a Montana Family's Hike Led to an Incredible Discovery." *Washington Post*, June 1, 2019. https://www.washingtonpost.com/local/the-nations-t-rex-how-a-montana-moms-hike-led-to-an-incredible-discovery/2019/06/01/2bd276f8-8252-11e9-bce7-40b4105f7ca0_story.html.

Henriques, Martha. "The Bitter Dinosaur Feud at the Heart of Palaeontology." BBC, January 19, 2023. https://www.bbc.com/future/article/20230119-the-dinosaur-feud-at-the-heart-of-palaeontology.

Henry, Thomas W., Steven M. Stanley, Karl W. Flessa, et al. "Society Records and Activities." *Journal of Paleontology* 70, no. 4 (1996): 702–11. http://www.jstor.org/stable/1306532.

Hodges, Montana. "Dinosaur Wars." *High Country News*, August 26, 2013. https://www.hcn.org/issues/45-14/dinosaur-wars/.

Holland, F. D., Jr., and Joseph H. Hartman. "Presentation of the Harrell L. Strimple Award of the Paleontological Society to Marshall E Lambert." *Journal of Paleontology* 70, no. 4 (July 1996): 708–9. https://doi.org/10.1017/s0022336000023702.

Horner, Jack. "An Intellectual Autobiography." *Montana Professor* 14, no. 2 (Spring 2004). https://mtprof.msun.edu/Spr2004/horner.html.

Horner, John R., and Edwin Dobb. *Dinosaur Lives: Unearthing an Evolutionary Saga*. HarperCollins, 1997.

Horner, John R., and James Gorman. *Digging Dinosaurs: The Search That Unraveled the Mystery of Baby Dinosaurs*. Workman, 1988.

———. *How to Build a Dinosaur: Extinction Doesn't Have to Be Forever*. Dutton, 2009.

Horner, J. R., and R. Makela. "Nest of Juveniles Provides Evidence of Family Structure Among Dinosaurs." *Nature* 282, no. 5736 (1979): 296–98. https://doi.org/10.1038/282296a0.

"How Do We Know Climate Change Is Real?" NASA Science, 2024. https://climate.nasa.gov/evidence/.

"How *T. rex* 'Lost' Its Third Claw." American Museum of Natural History, April 4, 2019. https://www.amnh.org/explore/news-blogs/on-exhibit-posts/t-rex-third-claw.

Inbody, Kristin. "Dinosaurs, the Bible, and a Glendive Museum." *Great Falls Tribune*, April 28, 2016.

"Introduction to Human Evolution." Smithsonian National Museum of Natural History. https://humanorigins.si.edu/education/introduction-human-evolution.

"Jack Horner, Paleontologist." Yale Center for Dyslexia and Creativity. https://dyslexia.yale.edu/story/jack-horner/.

Jaffe, Mark. *The Gilded Dinosaur: The Fossil War Between E. D. Cope and O. C. Marsh and the Rise of American Science*. Crown, 2000.

"The Jewel of the Prairie." Fort Peck Theatre. https://www.fortpecktheatre.org/theatre-history.

Johnson, Raymond N. "That Thou Canst Not Stir a Flower / Without Troubling of a Star." *Climate Science*, October 2018. https://www.icsusa.org/pages/articles/2018-icsusa-articles/october-2018---ldquothat-thou-canst-not-stir-a-flower-without-troubling-of-a-starrdquo-ndash-climate-and-man.php.

"Jordan, Montana." Wikipedia. https://en.wikipedia.org/wiki/Jordan,_Montana.

Kaiho, Kunio, Naga Oshima, Kouji Adachi, et al. "Global Climate Change Driven by Soot at the K-Pg Boundary as the Cause of the Mass Extinction." *Scientific Reports* 6, no. 1 (2016): 28427. https://doi.org/10.1038/srep28427.

Kaplan, Matt. "Dueling Dinosaurs Fizzle in Auction." *Nature*, November 19, 2013. https://www.nature.com/articles/nature.2013.14198.

Klotz, Irene. "Asteroid May Have Killed Dinosaurs Quicker than Scientists Thought." *Reuters*, February 8, 2013. https://www.reuters.com/article/us-space-asteroid-dinosaurs-idusbre91618a20130208/.

Koster, John. "Good the Old Bones: Dreaming of Dinosaurs, Digging for Dollars: Paleontologists Waged the 19th-Century 'Bone Wars' with Hired Hands. (Charles Hazelius Sternberg and Benjamin Franklin Mudge)." *Wild West* 25, no. 2 (2012): 26.

Kreier, Freda. "How Many *T. rex* Ever Existed? Calculation of Dinosaur's Abundance Offers an Answer." *Nature*, April 15, 2021. https://www.nature.com/articles/d41586-021-00984-2.

The Laurie Berkner Band. "We Are the Dinosaurs." 2017. https://open.spotify.com/track/21ryEt5nCkFjsHJHegnLOA?si=9dd546ec52684c0c.

Marin, Albro. *James J. Hill and the Opening of the Northwest*. Minnesota Historical Society Press, 1991.

Mayor, Adrienne. *Fossil Legends of the First Americans*. Princeton University Press, 2007. https://doi.org/10.1515/9781400849314.

McCarren, Mark J. *The Scientific Contributions of Othniel Charles Marsh: Birds, Bones, and Brontotheres*. Peabody Museum of Natural History, Yale University, 1993.

"Meet Our Team." Glendive Dinosaur and Fossil Museum. https://creationtruth.org/meet-our-team/.

"Mining for Mammoths in the Bad Lands." *New York Times*, December 3, 1905. https://timesmachine.nytimes.com/timesmachine/1905/12/03/119122159.html?pageNumber=29.

"Missouri River." Britannica. https://www.britannica.com/place/Missouri-River.

The Montana Dinosaur Trail. https://mtdinotrail.org/.

Montana Office of Tourism. "2021 Economic Impact." https://brand.mt.gov/_shared/Office-of-Tourism/docs/Fast-Facts-Funding-20b.pdf.

"Montana Paleontology Top Ten." Montana Dinosaur Trail, 2024. https://mtdinotrail.org/montana-paleontology-top-ten/.

"Montana Population, 1900–2023." Macrotrends. https://www.macrotrends.net/global-metrics/states/montana/population.

Moore, Kallie. *Tales of the Prehistoric World: Adventures from the Land of the Dinosaurs*. Neon Squid, 2022.

Moore, Sabre. "Rural Museums: Harnessing the Power of Place to Confront Silences and Revitalize Communities." PhD dissertation, Montana State University, 2023.

Murphy, Nate L., David Trexler, and Mark Thompson. "'Leonardo,' a Mummified Brachylophosaurus (Ornithischia: Hadrosauridae) from the Judith River Formation of Montana." In *Horns and Beaks: Ceratopsian and Ornithopod Dinosaurs*, edited by Kenneth Carpenter, 117–34. Indiana University Press, 2007. https://doi.org/10.2307/j.ctt1zxz1md.12.

"Museum of the Rockies." ProPublica. https://projects.propublica.org/nonprofits/organizations/816016828.

Nielson, Susie. "The 'Dueling Dinosaurs' Fossil Shows a *T. rex* and *Triceratops* in a Possible Fight. Researchers Are Now Poised to Unravel Its Mysteries." *Business Insider*, November 18, 2020. https://www.businessinsider.com/dueling-dinosaurs-fossil-t-rex-triceratops-bought-6-million-2020-11.

Osborn, Henry Fairfield. *Cope, Master Naturalist*. Arno Press, 1978.

Osterloff, Emily. "Dinosauria: How the 'Terrible Lizards' Got Their Name." Natural History Museum. https://www.nhm.ac.uk/discover/how-dinosaurs-got-their-name.html.

———. "The World's First Dinosaur Park: What the Victorians Got Right and Wrong." Natural History Museum. https://www.nhm.ac.uk/discover/crystal-palace-dinosaurs.html.

Pantuso, Phillip. "Perhaps the Best Dinosaur Fossil Ever Discovered. So Why Has Hardly Anyone Seen It?" *The Guardian*, July 17, 2019. https://www.theguardian.com/science/2019/jul/17/montana-fossilized-dueling-dinosaurs-skeletons-dino-cowboy.

Parsons, Chris. "A Moment That Changed Earth." *U.S. National Science Foundation*, June 15, 2022. https://new.nsf.gov/science-matters/moment-changed-earth.

Peppe, D. J., and A. L. Deino. "Dating Rocks and Fossils Using Geologic Methods." *Nature Education Knowledge* 4, no. 10 (2013): 1.

Preston, Douglas. "The Day the Dinosaurs Died." *New Yorker*, March 29, 2019. https://www.newyorker.com/magazine/2019/04/08/the-day-the-dinosaurs-died.

"Rainbow Falls (Missouri River)." Wikipedia. https://en.wikipedia.org/wiki/Rainbow_Falls_(Missouri_River).

Randall, David K. "He Was An All-Time Genius at Finding *Tyrannosaurus rexes*. His Story Will Break Your Heart." *Slate.com*, July 4, 2022. https://slate.com/news-and-politics/2022/07/t-rex-barnum-brown-dinosaur-collector-history.html.

———. *The Monster's Bones: The Discovery of T. rex and How It Shook Our World*. W. W. Norton, 2023.

———. "On Discovering the First Fossil of a *T. rex*." *Lit Hub*, June 10, 2022. https://lithub.com/on-discovering-the-first-fossil-of-a-t-rex/.

Renne, Paul R., Alan L. Deino, Frederik J. Hilgen, et al. "Time Scales of Critical Events Around the Cretaceous-Paleogene Boundary." *Science* 339, no. 6120 (2013): 684–87. https://doi.org/10.1126/science.1230492.

Sachar, Louis. *There's a Boy in the Girls' Bathroom*. Bullseye, 1987.

"Saving Peck's Rex." University of Notre Dame. https://www.nd.edu/stories/saving-pecks-rex/.

Schontzler, Gail. "Museum of the Rockies Adds $48 Million to Local Economy." *Bozeman Daily Chronicle*, September 17, 2014. https://www.bozemandailychronicle.com/news/education/museum-of-the-rockies-adds-48-million-to-local-economy/article_035d25ec-3eaa-11e4-8f68-4736aee6c30d.html.

Schuchert, Charles. *Biographical Memoir of Othniel Charles Marsh*. National Academy of Sciences, 1938.

Schweitzer, Mary. "Dinosaurs Offer a Rich Field for Study of the Human Era." *Scientific American*, June 2014. https://www.scientificamerican.com/article/dinosaurs-offer-a-rich-field-for-study-of-the-human-era/.

———. "Dinosaurs Reveal Clues about Adaptation to Climate." *Scientific American*, May 1, 2014. https://www.scientificamerican.com/article/dinosaurs-reveal-clues-about-adaptation-to-climate-change/.

*Secrets of the Dinosaur Mummy*. Directed by Michael Jorgenson. Myth Merchant Films, 2008.

Semonin, Paul. *American Monster: How the Nation's First Prehistoric Creature Became a Symbol of National Identity*. New York University Press, 2000.

Senter, Philip J. "Radiocarbon in Dinosaur Fossils: Compatibility with an Age of Millions of Years." *American Biology Teacher* 82, no. 2 (February 1, 2020): 72–79.

Society of Vertebrate Paleontology. "SVP Ethics Code." https://vertpaleo.org/code-of-conduct/.

Spears, Dean. "The World's Population May Peak in Your Lifetime. What Happens Next?" *New York Times*, September 18, 2023. https://www.nytimes.com/interactive/2023/09/18/opinion/human-population-global-growth.html.

Standing Rock Sioux Tribal Council. "Standing Rock Sioux Code of Justice Paleontology Resource Code." February 3, 2015. https://standingrock.org/wp-content/uploads/mdocs/Title%20xxxviii%20-%20(38)%20paleontology%20resource%20code.pdf.

"Standing Rock Sioux Tribe." North Dakota Native Tourism Alliance. https://www.ndnta.com/standingrocksioux.

Sternberg, Charles H. *The Life of a Fossil Hunter*. Indiana University Press, 1990.

Storrs, Glenn W. "*Elasmosaurus platyurus* and a Page from the Cope-Marsh war." *Discovery*, 1984. https://www.researchgate.net/publication/236201644_Elasmosaurus_platyurus_and_a_page_from_the_Cope-Marsh_War.

"These Small Town Students Sing and Dance Every Morning to Learn How to Get Along." *TODAY Show*, February 12, 2017. https://www.today.com/video/these-small-town-students-sing-and-dance-every-morning-to-learn-how-to-get-along-875545667775.

"The Town That Hunts Bones." *LIFE*, November 8, 1954.

Trexler, David. *Becoming Dinosaurs: A Prehistoric Perspective on Climate Change Today*. Sweetgrass Books, 2012.

———. "Marion Brandvold Celebrates 100th Birthday." *Choteau Acantha*, April 25, 2012. https://www.choteauacantha.com/news/article_eb1151ce-3ad5-5be2-ad09-412b48819a3f.html.

U.S. Army Corps of Engineers. "Historical Vignette." https://www.nwo.usace.army.mil/Media/Fact-Sheets/Fact-Sheet-Article-View/Article/562344/historical-vignette-fort-peck-dam/.

Viola, Herman. "Native Peoples, Lewis and Clark, and Mapmaking." Smithsonian Institution and Edgate.com, 2000. https://edgate.com/lewisandclark/indian_country.html.

Visit Rapid City Staff. "How a Depression-Era Park Became a Favorite Rapid City Attraction." Visit Rapid City, 22 May 2024, https://www.visitrapidcity.com/blog/post/how-a-depression-era-park-became-a-favorite-rapid-city-attraction/.

Wallace, David Rains. *The Bonehunters' Revenge: Dinosaurs, Greed, and the Greatest Scientific Feud of the Gilded Age*. Houghton Mifflin, 1999.

Water Science School. "The Atmosphere and the Water Cycle." U.S. Geological Survey, June 8, 2019. https://www.usgs.gov/special-topics/water-science-school/science/atmosphere-and-water-cycle.

"Welcome to Dinosaur Country!" Montana Department of Transportation. https://www.mdt.mt.gov/travinfo/docs/roadsigns/DinosaurCountry.pdf.

"What Does It Mean To Be Human? *Sahelanthropus tchadensis*." Smithsonian National Museum of Natural History. https://humanorigins.si.edu/evidence/human-fossils/species/sahelanthropus-tchadensis.

"What Is a Paleontologist?" American Geosciences Institute. https://www.americangeosciences.org/education/k5geosource/careers/paleontologist.